Muscle Cells - Recent Advances and Future Perspectives

Edited by Mani T. Valarmathi

Published in London, United Kingdom

Supporting open minds since 2005

Muscle Cells - Recent Advances and Future Perspectives
http://dx.doi.org/10.5772/intechopen.77689
Edited by Mani T. Valarmathi

Contributors
Miguel Luiz Batista Júnior, Felipe Henriques, Maria Cristina Cintra Gomes-Marcondes, Lais Rosa Viana, Andre Gustavo Oliveira, Bread Cruz, Rafael Rossi Valentim, Luiz Alberto Ferreira Ramos, Natalia Angelo Da Silva Miyaguti, Sarah Christine Pereira De Oliveira, Alexandra V. Ulyanova, Kentu Lassiter, Sami Dridi, Tiago Fernandes, Noemy Pereira, Camila Gatto, Edilamar Menezes de Oliveira, Kamal Ranjbar, Bayan Fayazi, Doris Hissako Sumida, Fernando Yamamoto Chiba, Maria Sara De Lima Coutinho Mattera, Gisela Gaina

First published in London, United Kingdom, 2020 by IntechOpen
IntechOpen is the global imprint of INTECHOPEN LIMITED, registered in England and Wales, registration number: 11086078, 7th floor, 10 Lower Thames Street, London, EC3R 6AF, United Kingdom
Printed in Croatia

British Library Cataloguing-in-Publication Data
A catalogue record for this book is available from the British Library

Additional hard and PDF copies can be obtained from orders@intechopen.com

Muscle Cells - Recent Advances and Future Perspectives
Edited by Mani T. Valarmathi
p. cm.
Print ISBN 978-1-78923-967-6
Online ISBN 978-1-78923-968-3
eBook (PDF) ISBN 978-1-83968-010-6

For EU product safety concerns:
IN TECH d.o.o., Prolaz Marije Krucifikse Kozulić 3, 51000 Rijeka, Croatia, info@intechopen.com or visit our website at intechopen.com.

Meet the editor

Mani T. Valarmathi is presently an assistant professor at the University of Alabama at Birmingham, USA. He began his scientific career as a cancer geneticist, but soon became captivated with the emerging and translational fields of stem cell biology, tissue engineering, and regenerative medicine. After completing his Bachelor's degree in Chemistry at the University of Madras, he received his MBBS in Medicine and Surgery and MD in Pathology from the University of Madras, as well as his PhD in Medical Biotechnology from All-India Institute of Medical Sciences, New Delhi, India. For over 15 years, he has had extensive experience in research on various types of stem cells, including adult, embryonic (pluripotent), and induced pluripotent stem cells. Currently, his research work focuses on generating three-dimensional vascularized tissues and/or organs for implantation purposes. He is a member of many prestigious national and international professional societies and scientific organizations, such as ISSCR, TERMIS, AACR, ASIP, ACS, ESC, ISHR, ASGCT, and AHA.

Contents

Preface XI

Section 1
Muscle Function: Cellular-Molecular 1

Chapter 1 3
Vascularisation of Skeletal Muscle
by Kamal Ranjbar and Bayan Fayazi

Chapter 2 27
Excitability of Vascular Smooth Muscle
by Alexandra V. Ulyanova

Chapter 3 41
Orexin System and Avian Muscle Mitochondria
by Kentu Lassiter and Sami Dridi

Chapter 4 59
Noncoding RNAs in the Cardiovascular System: Exercise Training Effects
by Noemy Pereira, Camila Gatto, Edilamar Menezes de Oliveira and Tiago Fernandes

Section 2
Muscle Disorder: Inflammation-Induced 85

Chapter 5 87
Inflammatory Muscle Diseases
by Doris Hissako Sumida, Fernando Yamamoto Chiba and Maria Sara de Lima Coutinho Mattera

Section 3
Muscle Wasting: Cancer-Induced 107

Chapter 6 109
Leucine and Its Importance for Cell Signalling Pathways in Cancer Cachexia-Induced Muscle Wasting
by Andre Gustavo Oliveira, Bread Cruz, Sarah Christine Pereira de Oliveira, Lais Rosa Viana, Natalia Angelo Da Silva Miyaguti, Luiz Alberto Ferreira Ramos, Rafael Rossi Valentim and Maria Cristina Cintra Gomes-Marcondes

Chapter 7 **131**
Adipose Tissue Remodeling during Cancer Cachexia
by Miguel Luiz Batista Júnior and Felipe Henriques

Section 4
Muscle Markers: Immuno-Analysis **147**

Chapter 8 **149**
Current Approaches in Immunoassay Methods Focus on Skeletal Muscle Proteins
by Gisela Gaina

Preface

The three different types of muscle tissue found in the animal kingdom are cardiac, skeletal, and smooth. The muscle cells are not only complex but also fascinating. In recent years there has been substantial advances in our understanding of muscle cell biology, especially in the areas of molecular anatomy, basic physiology, understanding disease mechanisms, and therapeutic targets.

Consequently, this book mainly focuses not only on the biology of myocytes, but also on all-encompassing disciplines pertaining to muscle tissue, such as fundamental physiology, molecular mechanisms of diseases, muscle regeneration, etc. for all three types of muscle, namely, skeletal, cardiac, and smooth muscle. As a result, the goal of this book is to consolidate the recent advances in the areas of muscle biology/diseases/regeneration covering a broad range of interrelated topics in a timely fashion and to disseminate that knowledge in a lucid way to a greater scientific audience.

The book consists of eight chapters, contributed by leading experts in basic science and clinical care, and is organized into four sections. The first section introduces the cellular and molecular aspects of muscle function, such as the role of vascularization during skeletal muscle regenerative and/or reparative processes, electrophysiological signature of arteriolar smooth muscle in various tissues with an emphasis on the excitability of fourth-order arterioles of skeletal muscle, the expression pattern and functional role of two orexigenic peptides (orexin system) in muscle mitochondrial dynamics, and finally the role of non-coding RNAs in exercise.

The second section of the book deals with current perspectives of inflammatory myopathies. Next, the third section explores the cancer-induced muscle-wasting disorder, focusing on the anabolic effects of branched-chain amino acids supplementation, especially leucine on skeletal muscle wasting and thereby ameliorating the host response to cancer-cachexia-induced muscle damage, as well as repercussions of adipose tissue dysfunction and remodeling and its significance in the development of cachexia. Eventually, the last section of the book contains a survey of recent advances in immunological detection assays pertaining to skeletal muscle proteins in physiological and pathological conditions.

This book will prove highly useful for students, researchers, and clinicians in muscle cell biology, exercise physiology/science, stem cell biology, developmental biology, cancer biology, pathology, oncology, as well as tissue engineering and regenerative medicine. This quick reference will benefit anyone desiring a thorough knowledge pertaining to recent advances in muscle biology in the context of health and disease.

I would like to thank the staff of IntechOpen who have produced this book so efficiently, and in particular I am indebted to Sandra Maljavac, the Author Service Manager, and Petra Svob, the Commissioning Editor, for their valuable source of advice throughout the preparation of this book. Finally, this book is dedicated to the memory of my parents and to the memory of my eldest brother.

Mani T. Valarmathi, MD, PhD
Department of Biomedical Engineering,
School of Medicine and School of Engineering,
UAB | The University of Alabama at Birmingham,
Birmingham, Alabama, USA

Section 1

Muscle Function: Cellular-Molecular

Chapter 1

Vascularisation of Skeletal Muscle

Kamal Ranjbar and Bayan Fayazi

Abstract

Skeletal muscle is mainly involved in physical activity and movement, which requires a large amount of glucose, fatty acids, and oxygen. These materials are supplied by blood vessels and incorporated into the muscle fiber through the cell membrane. In contrast, metabolic waste is discarded outside the cell membrane and removed by blood vessels. The formation of a functional, integrated vascular network is a fundamental process in the growth and maintenance of skeletal muscle. On the other hand, vascularization is one of the main central components in skeletal muscle regeneration. In order for regeneration to occur, blood vessels must invade the transplanted muscle. This is confirmed by the fact that muscle regeneration occurred from the outside of the muscle bundle toward the inner regions. In fact, it is likely that capillary formation is a key process to start muscle regeneration. Thus, vascularization activates muscle regeneration, and a decrease in vascularization could lead to disruption the process of muscle regeneration. Also, a better understanding of vascularization of skeletal muscle necessary for the successful formation of collateral arteries and recovery of injured skeletal muscle may lead to more successful strategies for skeletal muscle regeneration and engineering. So, in this chapter, we want to review vascularization in skeletal muscle.

Keywords: vascularization, angiogenesis, arteriogenesis, skeletal muscle

1. Introduction

Vascularization of skeletal muscle occurs by four distinct processes: vasculogenesis, angiogenesis, arteriogenesis, and lymphogenesis.

1.1 Vasculogenesis

The fundamental biological phenomenon of new vessel formation is one enormous complexity, and has excited the interest of scientific workers for many years. Vasculogenesis describes the formation of the primitive network of blood vessel in the embryo from undifferentiated precursor cells (angioblasts), and the differentiation of angioblasts to endothelial cell. This is the initial step to blood vessel formation de novo. To form a new vessel, angioblasts proliferate and join up with primary capillary plexus. The endothelial cell grid manufactured by vasculogenesis then serves as an angiogenesis framework [1]. After primary capillary plexus formation, it is altered by the sprouting and branching of new vessels from pre-existing ones. The majority of work on skeletal muscle capillaries has focused on angiogenesis [1].

1.2 Angiogenesis

Angiogenesis and inflammation are critical to the process of muscle regeneration, but the complex interactions between these multiple cell types are poorly understood [2]. Most normal angiogenesis occurs in the embryo where it establishes the primary vascular tree as well as an adequate vasculature for growing and developing skeletal muscles. Angiogenesis involves the growth of new capillaries from existing blood vessels within the skeletal muscle and occurs in the adult during the ovarian cycle and in physiological repair processes such as wound healing. However, very little turnover of endothelial cells occurs in the adult vasculature. In order for new blood vessel sprouts to form, as previously described by Papetti et al. [1], mural cells (pericytes) must first be removed from the branching vessel. Endothelial cell basement membrane and extracellular matrix is then degraded and remodeled by specific proteases such as matrix metalloproteinase (MMPs), and the new matrix synthesized by stromal cells is then laid down. This new matrix, coupled with soluble growth factors, fosters the migration and proliferation of endothelial cells. After sufficient endothelial cell division has occurred, endothelial cells arrest in a monolayer and form a tube-like structure. Mural cells (pericytes in the microvasculature and smooth muscle cells in larger vessels) are recruited to the abluminal surface of the endothelium, and vessels uncovered by pericytes regress. Blood flow is then established in the new vessel [1]. Angiogenesis in the skeletal muscle can occur by two primary mechanisms: sprouting and intussusception.

1.2.1 Sprouting angiogenesis

Sprouting in angiogenesis refers to activated endothelial cells diverging from the existing vasculature, continuing through the encompassing matrix to form a cord-like structure (see **Figure 1**). The endothelial cell cord is changed into a tube and sticks to the extracellular matrix. It should be noted that the newly formed tube must reenter the capillary network via joining with another capillary or venule to become a functional capillary. Newly formed capillary in the beginning is leaky, but it blossoms to that of the original capillary when pericytes surround completely the endothelial cells. Like intussusception, sprouting needs to activate endothelial cells [3].

1.2.2 Capillary intussusceptions

Intussusceptions point to the process by which a mature capillary is divided into two separate capillaries from within, by the formation of a pillar-like structure or longitudinal divide on the luminal side of the capillary (**Figure 2**). Activated endothelial cells extend intraluminally, effectively forming two tubes through which blood can pass. Previous studies confirmed that intussusception is a main method of capillary formation during development. Angiogenic responses were differential in vivo. In this case, it is mentioned that shear forces acting on capillaries may preferentially increase capillarity through intussusception [3]. On the other hand, as shown in **Figure 3**, overload by activation of MMPs and VEGF leads to angiogenesis via sprouting.

1.3 Arteriogenesis

Arteriogenesis, formerly regarded as a variant of angiogenesis, is a relatively new term that was introduced to distinguish it from other mechanisms of vascular growth, such as angiogenesis and vasculogenesis. Vasculogenesis describes the

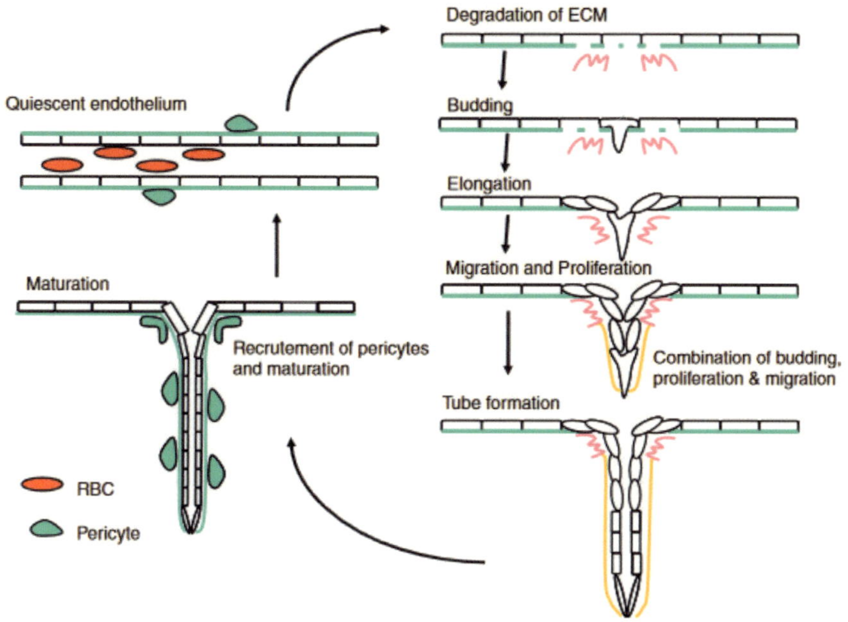

Figure 1.
Schematic representation of sprouting angiogenesis [4].

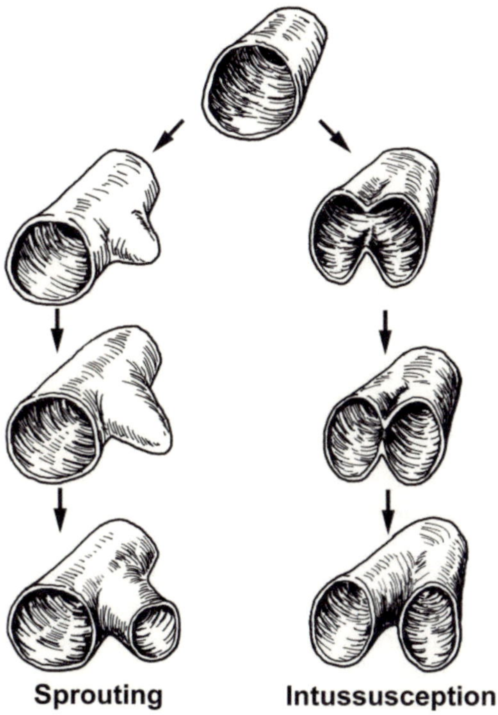

Figure 2.
Angiogenesis occurs by the processes of intussusception and sprouting.

embryonic development of blood vessels from angioblasts, and angiogenesis is the formation of new capillaries by sprouting and intussusception from pre-existent capillaries. But, arteriogenesis describes the formation of mature arteries from pre-existent interconnecting arterioles after an arterial occlusion. Arteriogenesis pointed

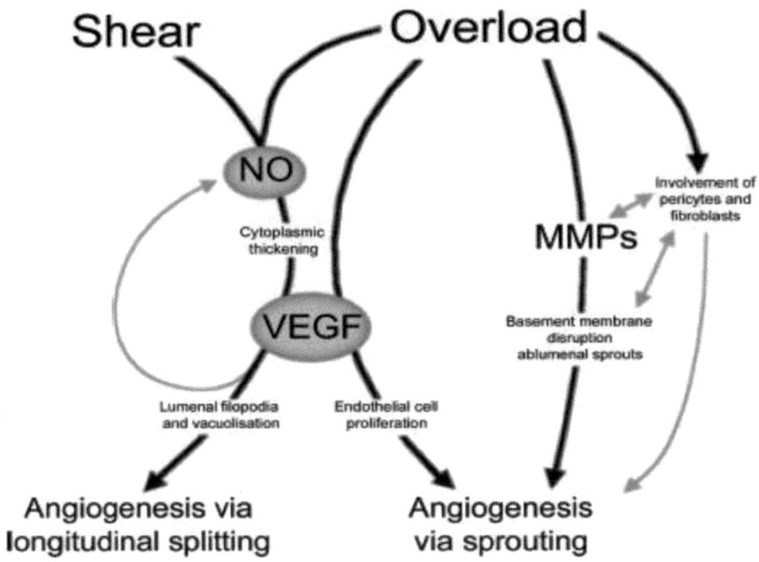

Figure 3.
Differential angiogenic responses in vivo [5].

to the enlargement of existing arterial vessels [6]. This enlargement indicates an increase in the caliber (diameter) and wall dimensions, resulting in a larger vessel. It is similar to angiogenesis in some features, but the pathways make it different [7].

Angiogenesis occurs under hypoxia/ischemia condition, which leads to the activation of the transcription factor HIF (hypoxia-inducible factor); in contrast, arteriogenesis is activated in an environment of normoxia [8].

Growth of arterioles in skeletal muscle has been documented during development.

Transformation of pericytes into smooth muscle cells and/or apposition of mesenchymal cells, most likely fibroblasts, to the abluminal surface of capillaries, followed by their gradual change into pericytes or smooth muscle cells lead to the growth of arteriolar known as "arteriolarization" [9]. Arteriolar growth accompanied rather than followed capillary growth.

1.4 Lymphogenesis

Lymphatic vessels act in close relation with blood vessels. The formed lymph fluid is transported via initial lymphatic capillaries to collecting vessels, to lymph nodes, and finally back to the blood [10]. Hyperemia-induced increased filtration of skeletal muscle (during activity), promotes hydrostatic and colloid osmotic pressure and addresses the need for increased lymph flow to maintain optimal conditions in the muscle. It is clear that exercise increases skeletal muscle lymph flow significantly in both animals [11] and humans [12], especially at the beginning of exercise. Studies showed that skeletal muscles contain small capillary-sized lymphatic vessels, which are located next to blood capillaries between muscle fibers but are much fewer in number.

2. Stimulators of vascularization in skeletal muscle

The studies showed that hypoxia, shear stress, adenosine, and muscle stretch are the most important stimulators of the angiogenesis process which are more fully described.

2.1 Hypoxia

Tissue hypoxia is thought to upregulate a series of local factors that contribute to angiogenesis. The status of tissue oxygenation determines blood vessels to undergo angiogenesis or stay quiescent. Accumulation of hypoxia-inducible factor (HIF)-α under hypoxia condition triggers tissue angiogenesis. HIF-α plays a pivotal role in the transcriptional activation of genes encoding angiogenic factors. Briefly, as previously described by Fong et al. [13], HIF-α abundance is negatively regulated by a subfamily of deoxygenates referred to as prolyl hydroxylase domain-containing proteins (PHDs), which use O_2 as a substrate to hydroxylate HIF-α subunits and hence tag them for rapid degradation (**Figure 4**). Under hypoxic conditions, HIF-α subunits accumulate due to reduced hydroxylation efficiency and form transcriptionally active heterodimers with HIF-1β to activate the expression of angiogenic factors and other proteins important for cellular adaptation to hypoxia. Angiogenesis is regulated by a combination of at least two different mechanisms. The paracrine mechanism is mediated by nonendothelial expression of angiogenic factors such as vascular endothelial growth factor (VEGF)-A, which in turn interacts with endothelial cell surface receptors to initiate angiogenic activities. In the autocrine mechanism, endothelial cells themselves are induced to express VEGF-A, which collaborate with the paracrine mechanism to support angiogenesis and protect vascular integrity [13].

2.2 Shear stress

Relatively little is known about the importance of mechanical forces and the mechanisms of their transduction during the growth of vessels in skeletal muscle [14]. Repeated muscle contractions alter the local microcirculatory hemodynamics by dilatation of arterioles leading to increases in capillary flow velocity, shear stress (defined as shear stress = blood viscosity × (8 × mean flow velocity)/vessel diameter) [15] and, potentially, pressure. Capillary growth in response to shear stress proceeds by division of the lumen by endothelial cell protrusion and vessel splitting, without the requirement for disturbance and breakdown of the basement membrane [14].

According to the mechanotransduction hypothesis, shear-induced endothelial cell deformation, cytoskeletal perturbations, and increasing tension at the focal adhesion attachment sites activate the integrins, generating intracellular signals that include transcription of genes involved in angiogenesis [15].

It is known that increased shear stress in endothelial cell cultures leads to an increase in protein expression of VEGF receptor 2 (Flk-1) [16]. Today it is clear that shear stress releases NO and increases VEGF and its receptor 2 expression during the initiation of endothelial cell proliferation and angiogenesis [14].

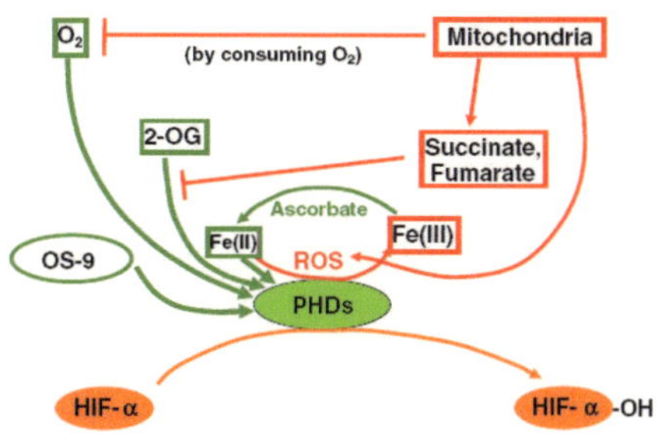

Figure 4.
Regulatory mechanisms of PHD hydroxylase activities. Factors or processes with positive effects on PHD hydroxylase activities are shown in green, whereas those with inhibitory effects are shown in red [13].

2.3 Adenosine

Of the major metabolites produced and released by exercising skeletal muscle, adenosine has received the greatest attention as an angiogenic factor. Adenosine is generated as ATP is catabolized when energy demands increase or oxygen supply decreases [17]. Feoktistov et al. demonstrated that stimulation of adenosine A_2B receptors upregulates the angiogenic factors VEGF and IL-8 in human endothelial cells under normoxic conditions [18]. Studies in other cell culture models conducted under normoxic conditions have also indicated that adenosine upregulates proangiogenic and downregulates antiangiogenic factors [19]. Remarkably, chronic infusion of adenosine induced neovascularisation in the skeletal muscle and the heart muscle [20].

2.4 Muscle stretch and exercise

Original studies demonstrated that stretch of cells promotes VEGF (mRNA and protein), which leads to enhanced endothelial cell migration and tube formation, and activates MT1-MMP, and upregulates Ang-2 and Tie-2 expression. Stretch also causes deformation of the extracellular matrix, and may result in the release of matrix-bound growth factors, that then can bind to and activate the surrounding cells. Furthermore, stretch induces tensional forces that initiate mechanotransduction signaling pathways through activation of integrin receptors [15].

When specific muscles are exercise trained, there can be an increase in flow capacity and an expected increase in the caliber of the large conduit vessels [21]. In this regard, Hounker et al. demonstrated that the subclavian arteries of the dominant arms of elite tennis players exhibited larger diameters than control group arteries [22]. Also, increased diameters of the femoral arteries were observed in elite cyclists, whereas decreases were observed with paraplegia compared with corresponding controls [21]. This increase in flow capacity to the major muscle groups of highly trained individuals could contribute to the greater exercise capacity exhibited by these individuals. Capillarity in active skeletal muscle is significantly increased by endurance exercise training although the increase in the heart muscle is less well established.

In this regard, laboratory studies indicate muscle mass loss [23], increased number of fast-twitch fibers, decreased slow-twitch fibers [24], and restriction of skeletal muscle blood flow in humans [25] and animals [26] after myocardial infarction. In addition, systemic blood flow in the skeletal muscle of mice with heart failure decreases at rest and during physical activity [27]. Experimental studies at the capillary level show that capillary density parameters [28] and capillary/fiber ratio (CF ratio) decrease following myocardial infarction. On the other hand, at the arteriole level, the arteriole tonic sympathetic vasoconstriction activity increases that decreases the arteriolar diameter and elasticity capacity [29].

A remaining question is what signaling cascade within the muscle fibers decodes muscle contractile activity signals in regulating VEGF expression. Mechanistically, the functional role of PGC-1α in exercise-induced VEGF expression and angiogenesis is dependent on the upstream p38 mitogen-activated protein kinase (p38ɣ MAPK) [30] and the downstream ERRα [31]. Interestingly, transgenic mice with muscle-specific expression of an inactive form of 5′-adenosine monophosphate-activated protein kinase (AMPK) have lower capillarity compared with the wild-type littermates, but have normal-induced angiogenesis in response to voluntary running exercise [32, 33].

The importance of other growth factor pathways remains to be elucidated. Although several signaling pathways have been identified (**Figure 5**), significant

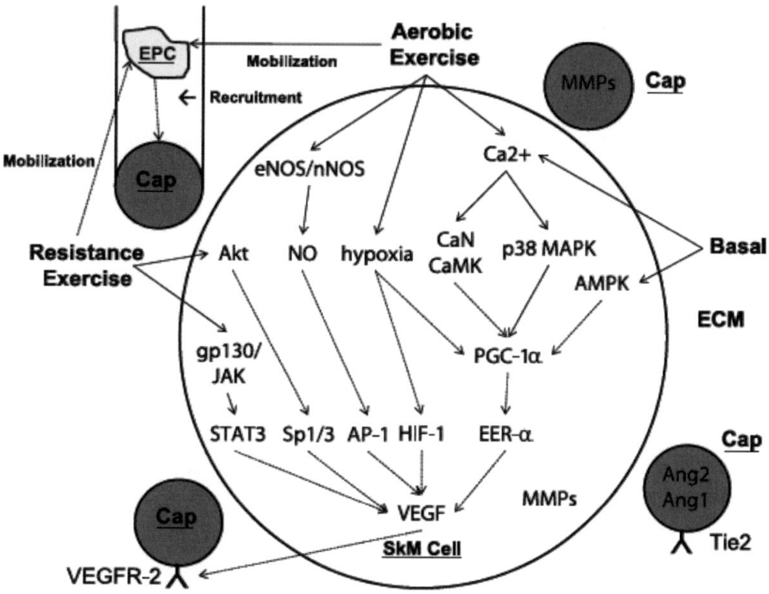

Figure 5.
Proposed model including intracellular signaling and growth factor regulation in skeletal muscle. Solid lines (—) are established pathways. Dashed lines (- -) are hypothesized pathways [34].

work remains on understanding the intracellular signaling pathways and transcription factors involved in basal- and exercise-induced angiogenesis [34].

It is well known that the vascularization of skeletal muscle adapts to various physiological and pathological conditions such as fiber type, gender, aging, obesity, and diabetes.

3. Fiber type and vascularization

The skeletal muscle is composed of a combination of different muscle fiber types: I, IIa, and IId/x. Fast- and slow-twitch fibers have different phenotypes with Type I fibers demonstrating the greatest and Type IId/x fibers demonstrating the least mitochondrial volume and capillarization.

It has been confirmed that muscle fibers with a high oxidative potential are related to a denser capillary network. This would mean that highly oxidative fibers require a higher rate of oxygen delivery than nonoxidative fibers; therefore, they must be better supplied with a capillary network [35].

Capillary density did not always correlate with oxidative capacity or maximal blood flow.

Previous studies demonstrated that after triiodothyronine administration, rat soleus and white area of the medial head of gastrocnemius muscle capillarity are promoted, whereas the oxidative capacity increased in the soleus only [36]. Therefore, in strait skeletal muscles, oxidative capacity is not the only factor that determines capillarity. On the other hand, it is likely that anaerobic waste accumulation stimulates vascularization in glycolytic regions of rat skeletal muscles, which is not accompanied by changes in oxidative metabolism [37, 38].

Already, at first, some researchers showed that capillary density is not related to the type of muscles, but hypothesized that the mean number of capillary profiles around a fiber for red and white muscles due to the size of the fiber. Sullivan and Pitman confirmed that capillaries around glycolytic fibers must supply blood stream to a greater volume of a muscle fiber than capillaries serving oxidative fibers [39]. In this regard, independent of fiber type in human muscles, there is a positive correlation between local capillary-to-fiber ratio (LCFR) and fiber area [40].

Cebasek and co-workers showed that the length of capillaries per unit fiber length was larger in Soleus (Slow muscle) than in Extensor Digitorum Longus (EDL, Fast muscle) muscle. On the other hand, these researchers showed that capillary length per unit fiber volume was larger in EDL muscle. There was no difference in the length of capillaries per unit fiber surface area between the two muscles. Oxidative and glycolytic fibers differ in the length of capillaries per unit fiber surface area. This parameter probably reflects the oxidative capacity of muscle fibers [41].

Ranjbar et al. showed that the pattern of vascularization in response to exercise training is different between fast- and slow-twitch muscles. We showed that the 10-week exercise training significantly increased capillary density and capillary-to-fiber ratio ($P < 0.05$) in slow-twitch muscle, but did not change fast-twitch muscle capillary density and capillary-to-fiber ratio. Furthermore, arteriolar density in fast-twitch muscle increased remarkably ($P < 0.05$) in response to training, but slow-twitch muscle arteriolar density did not change in response to exercise in chronic heart failure rats. HIF-1 increased ($P < 0.01$) but VEGF and FGF-2 mRNA did not change in slow-twitch muscle after training. In fast-twitch muscle, HIF-1 mRNA increased ($P < 0.05$), and VEGF and angiostatin decreased ($P < 0.01$) significantly after training. We concluded that endurance training ameliorates fast- and slow-twitch muscle revascularization nonuniformly in chronic heart failure rats by increasing capillary density in slow-twitch muscle and arteriolar density in fast-twitch muscle. The difference in revascularization at slow- and fast-twitch muscles may be induced by the difference in angiogenic and angiostatic gene expression response to endurance training [42].

In this regard, we showed that smaller arterioles decreased in cardiac after myocardial infarction. Aerobic training and l-arginine increased the number of cardiac arterioles with 11–25 and 26–50 μm diameters parallel to TGF-β overexpression. In gastrocnemius muscle, the number of arterioles/mm2 was only increased in the 11–25 μm in response to training with and without l-rginine parallel to angiostatin downregulation. Soleus arteriolar density with different sizes was not different between experimental groups. Results showed that 10 weeks aerobic exercise training and l-arginine supplementation promotes arteriogenesis of heart and gastrocnemius muscles parallel to overexpression of TGF-β and downregulation of angiostatin in myocardial infarction rats [43].

On the other hand, Panisello et al. showed that SO fibers are more sensitive to intermittent hypobaric hypoxia than both fast fiber types [44]. Furthermore, Murakami et al. showed that capillary-to-fiber ratio, Microvessel diameter, expression level of VEGF, and number of microvessels in the soleus were significantly higher than those in the EDL muscle [45].

In conclusion, capillary supply is evidently well adapted to different muscle fiber types; consequently, an average capillary supply of heterogeneous muscle depends on the muscle composition.

4. Gender

Numerous studies have reported on the physiological differences among genders, but the effect of gender on neovascularization of skeletal muscle is not yet clear, and research in this field is limited.

Robbins et al. showed that men had a greater capillary-to-fiber ratio than women. However, capillary density per square millimeter was not different between men and women [46]. Also, Kyriakides et al. showed that gender does not influence angiogenesis and arteriogenesis in the rabbit model of chronic hind limb ischemia [47]. Opposite to these findings, Keteyian et al. showed that capillary density at baseline was significantly greater in men (1.42 ± 0.08 endothelial cells × fiber $^{-1}$) than in women (1.12 ± 0.03 endothelial cells × fiber^{-1}) [48].

5. Aging

Aging effects on the structure and function of skeletal muscles have been intensely studied for decades; however, age-related changes in skeletal muscle capillarity still remain controversial [49].

Studies in both humans and animals have also produced conflicting results in skeletal muscle capillarity, with results in increases [50, 51], decreases [52, 53], or no change [54] in muscle capillaries with advancing age. These inconsistent reports could be due to a difference in factors, such as muscle type, fiber-type composition, and how this composition changes during aging, as well as to differences in subject activity levels and gender [55].

In general, it is believed that aging reduces the ability of an organism to respond to different types of stress [56]. For example, the angiogenic response to hind limb ischemia is impaired in aged compared with young mice [57, 58]. Researchers showed reduced angiogenic capacity in skeletal muscle with aging. Reduction of muscular oxidative capacity attainable through training in elderly people is related to the reduction in fitness that typically occurs in this population. The decrease in muscle blood flow in the aging period is related to altered reactivity of resistance arteries and arterioles, with impairment of both vasodilator and vasoconstrictor responses as a consequence of endothelial dysfunction [21]. Whilst aging and sedentary life style both lead to structural and functional impairment in skeletal muscle blood supply network, it is not yet clear whether impaired angiogenesis is an indirect response to reduced flow capacity. It is clear that reduction of VEGF secretion with maintaining of the ability to respond to exogenous cytokines is due to endothelial cell dysfunction with aging [21].

6. Obesity

Obesity is a major health problem in the United States and many other developed countries. Several lines of evidence have demonstrated that the capillary density and oxidative capacity appear to be more strongly correlated with insulin sensitivity and body fatness than the prevalence of a specific fiber type [59].

Increasing adiposity is associated with lower skeletal muscle oxidative capacity and capillarization. Plasma insulin elevation does not result in insulin increment in the interstitial space concentration between the capillary and muscle fiber [60], suggesting that a limit may exist in the diffusional conductance of insulin in skeletal muscle.

Research studies have shown that there are associations between lower insulin action and a lower muscle capillary-to-fiber area ratio in obese muscle. It is likely that compared with lean individual, a delayed transport of insulin over the capillary wall may be attributed to lower skeletal muscle capillary density in obese individuals [61].

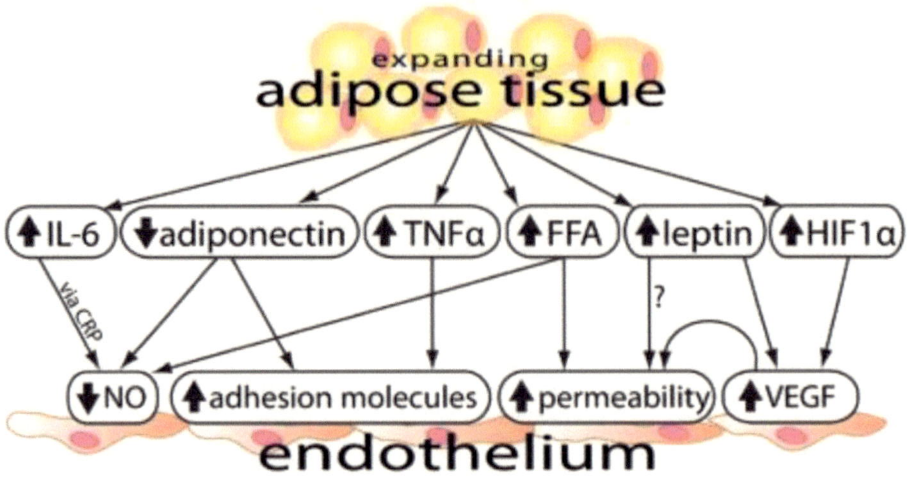

Figure 6.
Interactions between expanding adipose tissue and the endothelium [64].

In this regard, researchers showed that VEGF circulation, myocardial VEGF expression, and VEGF receptor 2 [kinase insert domain-containing receptor (KDR) human/Flk-1 murine analog] were not different between lean and obese individuals, but VEGF receptor 1 (Flt-1) is lower in obese compared with lean Zucker rats [62, 63]. Gavin et al. showed lower capillary density but no difference in VEGF expression in obese versus lean young skeletal muscle in humans [61].

The interactions between expanding adipose tissue and the endothelium via adipokine secretions are shown in **Figure 6** [64].

7. Diabetes

Another factor that affects the process of neovascularization in skeletal muscle is diabetes. Diabetes is a risk factor for peripheral vascular diseases, and it is associated with impaired collateral vessel growth in skeletal muscle. Diabetes, in turn, has been demonstrated to impair skeletal muscle and cardiac angiogenesis, and the mechanisms underlying this have generated much interest recently. Both type 1 and type 2 diabetes have been shown to affect angiogenic growth factors and inhibitors in skeletal muscle [65, 66].

Several proangiogenic protein gene expressions decreased and increased those of antiangiogenic ones in in diabetic mouse skeletal muscle [66]. Hence, the imbalance between stimulators and inhibitors may lead to peripheral cardiovascular complications in diabetes [67, 68].

8. Angiogenic factors

To better understand the neovascularization of skeletal muscle, it is important to understand the current state of knowledge regarding different factors possessing angiogenic properties. The process of angiogenesis in skeletal muscle is controlled by a number of factors that are released in the tissues surrounding the small vessel involved, and it is thought to be controlled by net balance between pro-angiogenic (angiogenic) and anti-angiogenesis (angiostatic) factors. In skeletal muscle, the balance of proangiogenic factors with antiangiogenic factors controls the extent of microvascular growth. For angiogenesis to occur, the balance of proangiogenic and antiangiogenic factors must favor the proangiogenic factors, and this has been termed the "angiogenic switch".

Therefore, a brief summary about the most well-known angiogenic (VEGF, FGF, and TGF) and angiostatic factors (endostatin, angiostatin, and thrombospondin-1) is presented here.

8.1 VEGF

Vascular endothelial growth factor (VEGF) is a 45 kDa secretable basic heparin-binding homodimeric glycoprotein. The human VEGF gene has been assigned to chromosome 6p21.3. Hypoxia is the main stimulus for VEGF production/expression. Observational studies support VEGF expression as an important factor for regulating skeletal muscle angiogenesis in both humans and animals [69, 70]. In patients with chronic disease conditions (e.g., chronic obstructive pulmonary disease, heart failure, and diabetes), as well as aging, locomotor skeletal muscle VEGF expression has also been reported to be lower, and these individuals often exhibit muscle weakness, reduced physical activity, and loss of skeletal muscle vascular density. A similar phenotype is found in sedentary (untrained) myocyte-specific VEGF gene ablated mice, which exhibit impaired exercise capacity and >50% loss of skeletal muscle microvessel density [71]. The human VEGF gene contains eight exons, seven introns, and a 14 kb coding region. VEGF has six different isoforms including VEGF165 (the predominant isoform), VEGF121, VEGF145, VEGF183, VEGF189, and VEGF 206 [72].

Binding of VEGF to the Flt-1 and KDR surface receptors activates their tyrosine kinase function resulting in enhanced endothelial cell proliferation, migration, vascular permeability, and protease activity [72, 73]. **Figure 7** shows several VEGF receptor signal transduction pathways that are known thus far.

8.2 FGFs

Currently, the family of FGFs consists of more than 20 members with 30.70% homology. FGFs are multifunctional proteins that bind to five cell membrane tyrosine kinase receptors (FGFR-1.5) and stimulate proliferation of a variety of cell types, including endothelial cells (ECs) and smooth muscle cells (SMCs). FGFs play a role in development, tissue regeneration, hematopoiesis, angiogenesis, and tumorigenesis. In vitro FGF-1, -2, -4, and -9 exert the highest mitogenic activity. Of these, FGF-1 and FGF-2 have previously been shown to induce therapeutic angiogenesis in vivo [75]. To the best of our knowledge, there are no published data about the potency of FGF-4 and FGF-9 in animal models [76].

8.3 Transforming growth factors (TGF-b) and platelet-derived growth factors (PDGF-BB)

Unlike VEGF and FGF, which are directly involved in the angiogenic process, TGF-b and PDGF-BB indirectly contribute to the angiogenic process. Mice lacking these two indicators die in utero due to defects in the process of vascular maturation. TGF and PDGF shear stress elements mediate upregulation of these factors in endothelial cells in response to increased shear stress in vitro.

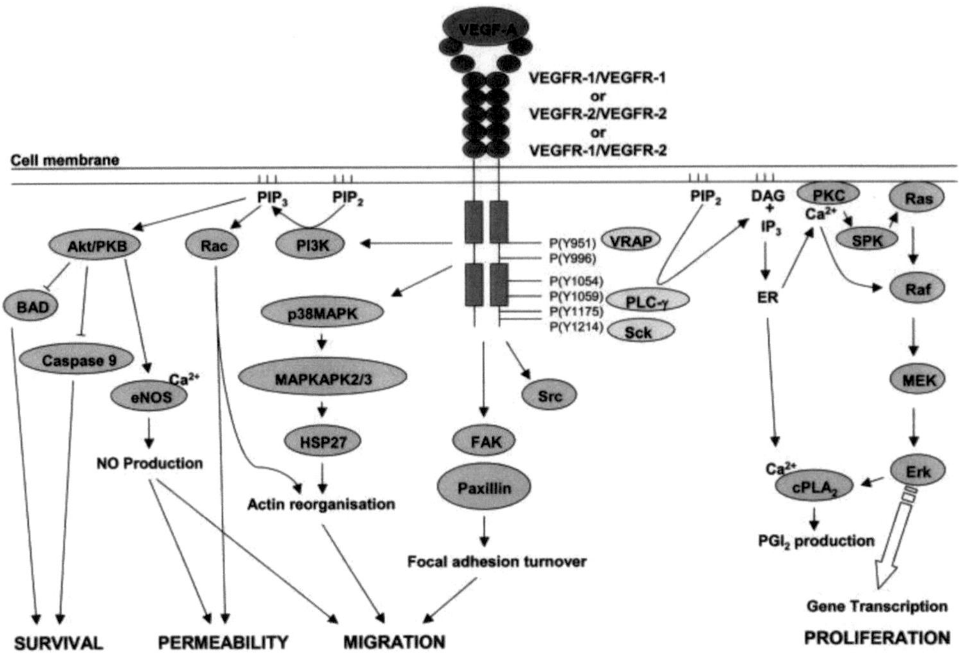

Figure 7.
VEGF receptor signal transduction. Several signal transduction pathways are activated by the binding of VEGF to its receptor, leading to increased proliferation, survival, permeability, and migration of cells [74].

PDGF also stimulates the proliferation of cultured smooth muscle cells and pericytes, both of which have been shown to express PDGF-b receptor [77].

TGF-b can control cell adhesion by regulating production of the extracellular matrix, stimulate or inhibit cell proliferation, protease inhibitors, and integrins, and induce cellular differentiation [78]. Much evidence points to an important role for TGF-b in the vasculature. It is clear that TGF-b recruits pericytes and smooth muscle cells in the arteriogenesis process [1].

8.4 Angiopoietin

During development and angiogenesis in adult tissue, there is a close relationship between a new group of factors, angiopoietins, and VEGF. The angiopoietins include angiopoietin 1 (Ang-1) and angiopoietin 2 (Ang-2). Similar to VEGF, angiopoietin receptors are located on the endothelial cells. For this reason, their function has been observed on endothelial cells. [79, 80]. Binding of Ang-1 to its receptors (Tie-2) maintains and stabilizes mature vessels by promoting interaction between the endothelial cells and surrounding cells [73]. In contrast, Ang-2 is thought to block the Tie-2 receptor and leads to vessel regression [73, 81]. It should to be noted that angiopoietins function depending on the VEGF present.

8.5 MMPs

Matrix metalloproteinases (MMPs) are a large family of protease enzymes and 26 members have been identified. Homeostasis of the extracellular matrix (ECM) in skeletal muscle dependent on MMP and TIMP balances [82]. The balance between angiogenic and anti-angiogenic factors interfere by MMPs via invoking angiogenic factors.

The ECM surrounds muscle fibers. ECM supports and protects muscle fiber and has a pivotal role in maintaining functional integrity of the fibers. Increased or decreased contractile activity of skeletal muscle promotes remodeling of the ECM. ECM degradation is a key step in the process of angiogenesis. Endothelial cells, stimulated by growth factors, produce MMPs that break down the cell membrane in physiological PH. Evidence is available that refers to the role of MMPs in the separation of smooth muscle cells from the extracellular matrix, and this allows the migration to cells [83].

Although several MMPs are found in skeletal muscle, mainly MMP-2 and MMP-9 play a more important role in skeletal muscle. MMP-2 and MMP-9 play an important role in skeletal muscle adaptation to changing contractile demands and to response to injury. MMP-2 and MMP-9 degrade type IV collagen and have important functions in homeostasis of the ECM during morphogenesis, proliferation, and cell apoptosis in a wide range of tissues. Expression of MMP-2 in skeletal muscle was increased following administration of chronic electrical stimulation, and expression of MMP-9 was increased after exposure to a chronic increase in blood flow [84].

8.6 Integrins

The communication between endothelial cells and the surrounding tissue (extracellular matrix: ECM) is affected by stretching via various mechanisms. Integrins are heterodimeric cell-surface receptors that link the cytoskeleton to the ECM; therefore, according to their status, integrins are involved in the vascularization process. One member of the integrin family, integrin αvβ3, is expressed on the surface of newly formed cells but is barely detectable in mature vessels. Differential expression of the alpha v family members may play a role in EC migration and apoptosis, though their role appears to be more complex than at first thought due to bidirectional signaling properties. Furthermore, interfering with integrin $\alpha_v\beta_3$ induces programmed cell death (apoptosis) in proliferating endothelial cells, which suggests its importance for the angiogenic process [80, 85, 86].

8.7 Follistatin-like 1

Follistatin-like 1 (Fstl1), also referred to as TSC36, is an extracellular glycoprotein that, despite limited homology, has been grouped into the follistatin family of proteins. Fstl1 is poorly understood with regard to its functional significance.

Studies indicated that Fstl1 is a secreted muscle protein or myokine that can function to promote endothelial cell function and stimulates revascularization in response to ischemic insult through its ability to activate Akt eNOS signaling [87].

8.8 Angiomotin

Angiomotin was recently identified as a new pro-angiogenic molecule. Angiomotin was detected at the surface of blood vessels of both healthy and pathological tissues such as placenta, retina, Kaposi's sarcoma, and breast tumors. Alternative splicing of angiomotin mRNA results in two protein isoforms of 80 and 130 kDa that exert very distinct roles during angiogenesis [88]. The p80 angiomotin isoform strongly stimulates *in vitro* and *in vivo* the migration of endothelial cells, a key event of the angiogenic process. Interestingly, angiostatin binding to angiomotin extracellular domain strongly inhibits such an effect. In contrast, p130 angiomotin isoform only exerts a very weak stimulatory effect on endothelial cell migration. The p130 protein was identified as tightly associated to cytoskeleton actin filaments and highly involved in vessel stabilization and maturation [89]. Interestingly, skeletal muscles represent the most abundant tissue of the body and they express high levels of angiomotin. Moreover, microcirculation is a critical component of muscle function since capillaries provide myofibers with oxygen and nutriments, and remove carbon dioxide and metabolic waste. As myofibers respond to physiological or pathological conditions with a remarkable plasticity, it is crucial that the microcirculation remains well matched with the myofibers' needs in order to preserve muscle function. Depending on conditions, such muscle angio-adaptation can involve either angiogenesis or some vascular regression. Given its role not only during angiogenesis but also for vessel stabilization and maturation, angiomotin might thus represent an important factor in muscle angio-adaptation. To date, angiomotin expression in skeletal muscle has never been investigated in response to physiological or pathological conditionings [89].

8.9 EPH-B4/EPHRIN-B2

A unique class of receptor/ligand pair, Eph receptors and ephrin ligands, plays a prominent role in blood vessel development. Interestingly, not only does an ephrin expressed on the surface of one cell bind and activate its cognate Eph receptor on another cell, but through a reciprocal signaling mechanism the ephrin is also activated upon receptor engagement [90].

Ephrin-B2 as a member of the ephrin family is explicit on arterial endothelial cells, and its receptor eph-B4 is localized to venous endothelial cells.

Interaction of ephrin-B2 and its receptor has determined the primary capillary plexus after vasculogenesis [91]. The exact role of this interaction in the angiogenic process is not completely clear. But what has been approved is that establishment of contact and signaling between arterial and venous compartments mediated by ephrin-B2 and eph-4B is necessary for remodeling of the established primary capillary plexus [1].

9. Angiostatic factors

Vascularization can be suppressed at any of a number of key steps in this process by endothelial growth cycle disruption in which a quiescent vessel becomes an actively growing and invading endothelial tube.

In this regard, vasculogenesis can be blocked by different factors. Degradation of the extracellular matrix by activated endothelial cells can also be prevented, which suppresses sprouting and invasion of growing capillaries into their surroundings. Alternatively, endothelial cell proliferation can be inhibited by agents that block signaling within the cell and arrest its division cycle or by agents that prevent the maturation of nascent endothelial cells into functional tubes .Finally, endothelial cells can be forced to apoptose, which destroys the existing vessels and thereby impairs survival of vessel and metastasis through the vascular route.

In addition to the numerous factors that stimulate angiogenesis, both physiologically and pathologically, many substances including those mentioned above can inhibit blood vessel growth.

9.1 Angiostatin

Angiostatin is an internal fragment of plasminogen. Angiostatin was the first proteolytic fragment described with anti-angiogenic activity, derived from plasminogen via MMP degradation of plasmin.

Inhibition of NOS increased the expression of angiostatin and activities of MMP-2 and MMP-9 which generate it, suggesting that compromised NO production may lead to impaired angiogenesis during endothelial dysfunction. Whether there is any role in modulating flow-mediated angiogenesis is unknown. Four potential "receptors" for angiostatin have been identified: integrin $\alpha_v\beta_3$, ATP synthase, angiomotin, and the NG2 chondroitin sulfate proteoglycan (CSPG).

9.2 Endostatin

Endostatin is a 20-kDa carboxyl terminal proteolytic cleavage fragment of collagen type XVIII [92]. It is thought to be generated through a two-step process, as follows: a metal-dependent early cleavage of collagen type XVIII, followed by cleavage at an Ala-His site. It is not entirely clear which protease(s) is responsible for endostatin generation from collagen type XVIII in vivo, as multiple proteases can cleave recombinant fragments of human collagen type XVIII to generate endostatin-like fragments. In a study testing approximately 12 different proteases, elastase and cathepsin-L were the only two proteases found to efficiently cleave fragments of recombinant human collagen type XVIII, generating endostatin-like fragments. In support of a role for cathepsin-L in the generation of endostatin, cathepsin-L can proteolytically generate endostatin from its precursor in murine hemangioendothelioma cells propagated in vitro. Endostatin was originally reported to inhibit the proliferation of bovine capillary endothelial cells, but not the proliferation of cells of nonendothelial origin, and to also inhibit angiogenesis in the chick chorioallantoic membrane model [92] . Endostatin receptors that mediate the biological effects of endostatin are not yet clear. In this regard, research studies have shown that endostatin has the ability to bind to various cell surface molecules such as glypican, integrin $\alpha5\beta1$, tropomyosin, and the VEGF receptor KDR/Flk-1 [93, 94].

9.3 Thrombospondin-1

Thrombospondin-1 (TSP-1) is a large (~450 kDa) multifunctional homotrimeric matrix glycoprotein whose action primarily serves to inhibit angiogenesis. Originally found to be stored and secreted in platelet α-granules, TSP-1 is now known to be produced by a wide variety of cells, including fibroblasts, keratinocytes, neutrophils, and macrophages and is believed to be a major secretory product of vascular smooth muscle and endothelial cells. The actions of TSP-1 include participation in platelet aggregation, inhibition of proteolytic enzymes, inhibition of endothelial cell proliferation, diminution of cell spreading, disruption of focal (cell-to-matrix) adhesions, and inhibition of angiogenesis *in vitro* and *in vivo*. However, the physiological relevance of TSP-1 in regulating skeletal muscle angiogenesis is not known. In these regard, Malek et al. showed that TSP-1 is an important endogenous negative regulator of angiogenesis that prevents excessive capillarization in the heart and skeletal muscles [95].

9.4 TIMPs

TIMPs inhibit MMP activities. Of the four TIMPs (TIMP-1, TIMP-2, TIMP-3, and TIMP-4) identified so far, TIMP not only has a direct effect on the growth and migration of endothelial cells, but also affects the extracellular matrix, which is an essential component of angiogenesis responses. The balance between stimulants and inhibitors of MMPs is a key step in the angiogenesis regulation. TIMP-1 and

TIMP-2 block the release of MMP-2 and MMP-9 zymogens, respectively. This factor reduces VEGF gene expression.

TIMP-1 is the most abundant type of TIMPs. TIMP-1 is a glycoprotein with a weight of 25.8 kDa and 184 amino acids, which has a variety of functions, such as growth factor activity, stimulation of morphological changes in cells, and prevention of angiogenesis. TIMP-1 with different gravities interacts with all known MMPs.

TIMP-2 is a nonglycolytic protein weighing 21 kDa and 196 amino acids, which prevents tumor growth. This protein mainly attaches to MMP-2. Of course, this factor, like TIMP-1, has the ability to connect to all MMPs.

TIMP-3 is a protein with a weight of 41 kDa and 188 amino acids, which causes cell death. This angiostatic factor specifically results in the deactivation of MMP-1, MMP-2, MMP3, and MMP-9. In this regard, TIMP-4 specifically connects to the MMP-2 and inhibits biological effects of MMP-2 [94].

9.5 Interferons

Interferons (INF-a, b, and g) are members of a family of secreted glycoproteins that were initially characterized for their antiviral effect. Interferons inhibit angiogenesis [96]. It is likely that IFN-a and IFN-b lead to downregulation of bFGF mRNA and protein levels [97] as well as its inhibitory effect on endothelial cell migration [1].

10. Conclusions

It is clear that a single paradigm cannot be used to encompass the unique patterns of capillary growth observed in response to various angiogenic stimuli. Over the past decade, novel markers of neovascularization in skeletal muscle have been identified, both at molecular and genetic levels, consequently leading our understanding of the molecular mechanisms involved in neovascularization of skeletal muscle to new heights. We have reviewed that the EC response during physiological angiogenesis within skeletal muscle is potentially sensitive to the hypoxia, shear stress, mechanical stimulus, and adenosine. A complex mix of humoral/metabolic and mechanical stimuli works coordinately within muscle to provide the cues that stimulate vascularization. The intricacies of these signaling pathways and levels of crosstalk between pathways remain to be elucidated. Although we still require delineation of signaling pathways evoked by individual angiogenic stimuli, a major goal for future research will be to determine how multiple stimuli are integrated within the capillary to determine a particular pattern of capillary growth. As mentioned, the process of vascularization in skeletal muscle depends on many factors such as angiogenic and angiostatic factors, but how their participation in the process of vascularization in the type of skeletal muscle fibers is not yet clear. Several key questions about the process of vascularization in skeletal muscle still remain to be addressed. For example: what is the relationship between intussusception, sprouting, angiogenic, and angiostatic factors and the type of skeletal muscle fibers? What is the relationship between angiogenesis factors secreted from muscle fibers and myosin heavy chains?

Author details

Kamal Ranjbar[1]* and Bayan Fayazi[2]

1 Department of Physical Education and Sport Science, Bandar Abbas Branch, Islamic Azad University, Bandar Abbas, Iran

2 Faculty of Physical Education and Sport Science, Razi University, Kermanshah, Iran

*Address all correspondence to: kamal_ranjbar2010@yahoo.com

References

[1] Papetti M, Herman IM. Mechanisms of normal and tumor-derived angiogenesis. American Journal of Physiology. Cell Physiology. 2002;**282**(5):C947-C970

[2] Shireman PK. The chemokine system in arteriogenesis and hind limb ischemia. Journal of Vascular Surgery. 2007;**45**(Suppl A):A48-A56

[3] Prior BM, Yang HT, Terjung RL. What makes vessels grow with exercise training? Journal of Applied Physiology. 2004;**97**(3):1119-1128

[4] Dufraine J, Funahashi Y, Kitajewski J. Notch signaling regulates tumor angiogenesis by diverse mechanisms. Oncogene. 2008;**27**(38):5132

[5] Egginton S. Physiological factors influencing capillary growth. Acta Physiologica (Oxford, England). 2011;**202**(3):225-239

[6] Egginton S. Invited review: Activity-induced angiogenesis. Pflügers Archiv. 2009;**457**(5):963-977

[7] Cai W, Schaper W. Mechanisms of arteriogenesis. Acta Biochimica et Biophysica Sinica Shanghai. 2008;**40**(8):681-692

[8] Helisch A et al. Impact of mouse strain differences in innate hindlimb collateral vasculature. Arteriosclerosis, Thrombosis, and Vascular Biology. 2006;**26**(3):520-526

[9] Hansen-Smith F et al. Growth of arterioles precedes that of capillaries in stretch-induced angiogenesis in skeletal muscle. Microvascular Research. 2001;**62**(1):1-14

[10] Kivela R et al. Effects of acute exercise, exercise training, and diabetes on the expression of lymphangiogenic growth factors and lymphatic vessels in skeletal muscle. American Journal of Physiology. Heart and Circulatory Physiology. 2007;**293**(4):H2573-H2579

[11] Coates G, O'Brodovich H, Goeree G. Hindlimb and lung lymph flows during prolonged exercise. Journal of Applied Physiology. 1993;**75**(2):633-638

[12] Havas E et al. Albumin clearance from human skeletal muscle during prolonged steady-state running. Experimental Physiology. 2000;**85**(6):863-868

[13] Fong GH. Regulation of angiogenesis by oxygen sensing mechanisms. Journal of Molecular Medicine. 2009;**87**(6):549-560

[14] Hudlicka O, Brown MD. Adaptation of skeletal muscle microvasculature to increased or decreased blood flow: Role of shear stress, nitric oxide and vascular endothelial growth factor. Journal of Vascular Research. 2009;**46**(5):504-512

[15] Haas TL. Molecular control of capillary growth in skeletal muscle. Canadian Journal of Applied Physiology. 2002;**27**(5):491-515

[16] Abumiya T et al. Shear stress induces expression of vascular endothelial growth factor receptor Flk-1/KDR through the CT-rich Sp1 binding site. Arteriosclerosis, Thrombosis, and Vascular Biology. 2002;**22**(6):907-913

[17] Fredholm BB et al. International Union of Pharmacology. XXV. Nomenclature and classification of adenosine receptors. Pharmacological Reviews. 2001;**53**(4):527-552

[18] Feoktistov I et al. Differential expression of adenosine receptors in human endothelial cells: Role of A2B receptors in angiogenic factor

regulation. Circulation Research. 2002;**90**(5):531-538

[19] Adair TH. Growth regulation of the vascular system: An emerging role for adenosine. American Journal of Physiology. Regulatory, Integrative and Comparative Physiology. 2005;**289**(2):R283-R296

[20] Ryzhov S et al. Role of adenosine receptors in the regulation of angiogenic factors and neovascularization in hypoxia. The Journal of Pharmacology and Experimental Therapeutics. 2007;**320**(2):565-572

[21] Bloor CM. Angiogenesis during exercise and training. Angiogenesis. 2005;**8**(3):263-271

[22] Huonker M et al. Size and blood flow of central and peripheral arteries in highly trained able-bodied and disabled athletes. Journal of Applied Physiology. 2003;**95**(2):685-691

[23] Magnusson G et al. Exercise capacity in heart failure patients: Relative importance of heart and skeletal muscle. Clinical Physiology. 1996;**16**(2):183-195

[24] Sullivan MJ, Green HJ, Cobb FR. Skeletal muscle biochemistry and histology in ambulatory patients with long-term heart failure. Circulation. 1990;**81**(2):518-527

[25] Drexler H et al. Alterations of skeletal muscle in chronic heart failure. Circulation. 1992;**85**(5):1751-1759

[26] Didion SP et al. Enhanced constrictor responses of skeletal muscle arterioles during chronic myocardial infarction. American Journal of Physiology - Heart and Circulatory Physiology. 1997;**273**(3):H1502-H1508

[27] Kindig CA et al. Impaired capillary hemodynamics in skeletal muscle of rats in chronic heart failure. Journal of Applied Physiology. 1999;**87**(2):652-660

[28] Schieffer B et al. Development and prevention of skeletal muscle structural alterations after experimental myocardial infarction. American Journal of Physiology - Heart and Circulatory Physiology. 1995;**269**(5):H1507-H1513

[29] Thomas DP, Hudlická O. Arteriolar reactivity and capillarization in chronically stimulated rat limb skeletal muscle post-MI. Journal of Applied Physiology. 1999;**87**(6):2259-2265

[30] Pogozelski AR et al. p38γ mitogen-activated protein kinase is a key regulator in skeletal muscle metabolic adaptation in mice. PLoS One. 2009;**4**(11):e7934

[31] Chinsomboon J et al. The transcriptional coactivator PGC-1alpha mediates exercise-induced angiogenesis in skeletal muscle. Proceedings of the National Academy of Sciences of the United States of America. 2009;**106**(50):21401-21406

[32] Yan Z et al. Regulation of exercise-induced fiber type transformation, mitochondrial biogenesis, and angiogenesis in skeletal muscle. Journal of Applied Physiology. 2011;**110**(1):264-274

[33] Zwetsloot KA et al. AMPK regulates basal skeletal muscle capillarization and VEGF expression, but is not necessary for the angiogenic response to exercise. The Journal of Physiology. 2008;**586**(Pt 24):6021-6035

[34] Gavin TP. Basal and exercise-induced regulation of skeletal muscle capillarization. Exercise and Sport Sciences Reviews. 2009;**37**(2):86-92

[35] Janacek J et al. 3D visualization and measurement of capillaries supplying metabolically different fiber types in the rat extensor digitorum longus muscle

during denervation and reinnervation. The Journal of Histochemistry and Cytochemistry. 2009;**57**(5):437-447

[36] Charifi N et al. Enhancement of microvessel tortuosity in the vastus lateralis muscle of old men in response to endurance training. The Journal of Physiology. 2004;**554**(Pt 2):559-569

[37] Badr I et al. Differences in local environment determine the site of physiological angiogenesis in rat skeletal muscle. Experimental Physiology. 2003;**88**(5):565-568

[38] Egginton S, Hudlicka O. Selective long-term electrical stimulation of fast glycolytic fibres increases capillary supply but not oxidative enzyme activity in rat skeletal muscles. Experimental Physiology. 2000;**85**(5):567-573

[39] Sullivan SM, Pittman RN. In vitro O2 uptake and histochemical fiber type of resting hamster muscles. Journal of Applied Physiology. 1984;**57**(1):246-253

[40] Ahmed SK et al. Is human skeletal muscle capillary supply modelled according to fibre size or fibre type? Experimental Physiology. 1997;**82**(1):231-234

[41] Cebasek V et al. Nerve injury affects the capillary supply in rat slow and fast muscles differently. Cell and Tissue Research. 2006;**323**(2):305-312

[42] Ranjbar K, Ardakanizade M, Nazem F. Endurance training induces fiber type-specific revascularization in hindlimb skeletal muscles of rats with chronic heart failure. Iranian journal of basic medical sciences. 2017;**20**(1):90

[43] Ranjbar K, Rahmani-Nia F, Shahabpour E. Aerobic training and l-arginine supplementation promotes rat heart and hindleg muscles arteriogenesis after myocardial infarction. Journal of Physiology and Biochemistry. 2016;**72**(3):393-404

[44] Panisello P et al. Capillary supply, fibre types and fibre morphometry in rat tibialis anterior and diaphragm muscles after intermittent exposure to hypobaric hypoxia. European Journal of Applied Physiology. 2008;**103**(2):203-213

[45] Murakami S et al. Comparison of capillary architecture between slow and fast muscles in rats using a confocal laser scanning microscope. Acta Medica Okayama. 2010;**64**(1):11-18

[46] Robbins JL et al. A sex-specific relationship between capillary density and anaerobic threshold. Journal of Applied Physiology. 2009;**106**(4):1181-1186

[47] Kyriakides ZS et al. Gender does not influence angiogenesis and arteriogenesis in a rabbit model of chronic hind limb ischemia. International Journal of Cardiology. 2003;**92**(1):83-91

[48] Keteyian SJ et al. Differential effects of exercise training in men and women with chronic heart failure. American Heart Journal. 2003;**145**(5):912-918

[49] Cui L et al. Arteriolar and venular capillary distribution in skeletal muscles of old rats. The Journals of Gerontology. Series A, Biological Sciences and Medical Sciences. 2008;**63**(9):928-935

[50] Davidson YS et al. The effect of aging on skeletal muscle capillarization in a murine model. The Journals of Gerontology. Series A, Biological Sciences and Medical Sciences. 1999;**54**(10):B448-B451

[51] Coggan AR et al. Histochemical and enzymatic characteristics of skeletal muscle in master athletes. Journal of Applied Physiology. 1990;**68**(5):1896-1901

[52] Degens H et al. Capillary proliferation related to fibre types in

hypertrophied aging rat M. Plantaris. Advances in Experimental Medicine and Biology. 1994;**345**:669-676

[53] Parizkova J et al. Body composition, aerobic capacity, and density of muscle capillaries in young and old men. Journal of Applied Physiology. 1971;**31**(3):323-325

[54] Kano Y et al. Effects of aging on capillary number and luminal size in rat soleus and plantaris muscles. The Journals of Gerontology. Series A, Biological Sciences and Medical Sciences. 2002;**57**(12):B422-B427

[55] Lyon MJ, Steer LM, Malmgren LT. Stereological estimates indicate that aging does not alter the capillary length density in the human posterior cricoarytenoid muscle. Journal of Applied Physiology. 2007;**103**(5):1815-1823

[56] Gavin TP et al. No difference in the skeletal muscle angiogenic response to aerobic exercise training between young and aged men. The Journal of Physiology. 2007;**585**(Pt 1):231-239

[57] Shimada T et al. Angiogenesis and vasculogenesis are impaired in the precocious-aging klotho mouse. Circulation. 2004;**110**(9):1148-1155

[58] Yu J et al. An engineered VEGF-activating zinc finger protein transcription factor improves blood flow and limb salvage in advanced-age mice. The FASEB Journal. 2006;**20**(3):479-481

[59] Shono N et al. Decreased skeletal muscle capillary density is related to higher serum levels of low-density lipoprotein cholesterol and apolipoprotein B in men. Metabolism. 1999;**48**(10):1267-1271

[60] Gudbjornsdottir S et al. Direct measurements of the permeability surface area for insulin and glucose in human skeletal muscle. The Journal of Clinical Endocrinology and Metabolism. 2003;**88**(10):4559-4564

[61] Gavin TP et al. Lower capillary density but no difference in VEGF expression in obese vs. lean young skeletal muscle in humans. Journal of Applied Physiology. 2005;**98**(1):315-321

[62] Chou E et al. Decreased cardiac expression of vascular endothelial growth factor and its receptors in insulin-resistant and diabetic states: A possible explanation for impaired collateral formation in cardiac tissue. Circulation. 2002;**105**(3):373-379

[63] Toblli JE et al. Angiotensin-converting enzyme inhibition and angiogenesis in myocardium of obese Zucker rats. American Journal of Hypertension. 2004;**17**(2):172-180

[64] Rutkowski JM, Davis KE, Scherer PE. Mechanisms of obesity and related pathologies: The macro- and microcirculation of adipose tissue. The FEBS Journal. 2009;**276**(20):5738-5746

[65] Galasso G et al. Impaired angiogenesis in glutathione peroxidase-1-deficient mice is associated with endothelial progenitor cell dysfunction. Circulation Research. 2006;**98**(2):254-261

[66] Kivela R et al. Effects of experimental type 1 diabetes and exercise training on angiogenic gene expression and capillarization in skeletal muscle. The FASEB Journal. 2006;**20**(9):1570-1572

[67] Kivela R et al. Exercise-induced expression of angiogenic growth factors in skeletal muscle and in capillaries of healthy and diabetic mice. Cardiovascular Diabetology. 2008;**7**:13

[68] Boodhwani M et al. Functional, cellular, and molecular characterization of the angiogenic response to chronic

myocardial ischemia in diabetes. Circulation. 2007;**116**(11 Suppl):I31-I37

[69] Olfert IM et al. Myocyte vascular endothelial growth factor is required for exercise-induced skeletal muscle angiogenesis. American Journal of Physiology. Regulatory, Integrative and Comparative Physiology. 2010;**299**(4):R1059-R1067

[70] Prior BM et al. Exercise-induced vascular remodeling. Exercise and Sport Sciences Reviews. 2003;**31**(1):26-33

[71] Olfert IM et al. Muscle-specific VEGF deficiency greatly reduces exercise endurance in mice. The Journal of Physiology. 2009;**587**(Pt 8):1755-1767

[72] Lockwood CJ, Schatz F, Krikun G. Angiogenic factors and the endometrium following long term progestin only contraception. Histology and Histopathology. 2004;**19**(1):167-172

[73] Qazi Y, Maddula S, Ambati BK. Mediators of ocular angiogenesis. Journal of Genetics. 2009;**88**(4):495-515

[74] Hoeben A et al. Vascular endothelial growth factor and angiogenesis. Pharmacological Reviews. 2004;**56**(4):549-580

[75] Muhlhauser J et al. In vivo angiogenesis induced by recombinant adenovirus vectors coding either for secreted or nonsecreted forms of acidic fibroblast growth factor. Human Gene Therapy. 1995;**6**(11):1457-1465

[76] Rissanen TT et al. Fibroblast growth factor 4 induces vascular permeability, angiogenesis and arteriogenesis in a rabbit hindlimb ischemia model. The FASEB Journal. 2003;**17**(1):100-102

[77] Nicosia RF, Nicosia SV, Smith M. Vascular endothelial growth factor, platelet-derived growth factor, and insulin-like growth factor-1 promote rat aortic angiogenesis in vitro. The American Journal of Pathology. 1994;**145**(5):1023-1029

[78] Massague J. The transforming growth factor-beta family. Annual Review of Cell Biology. 1990;**6**:597-641

[79] Sato TN et al. Distinct roles of the receptor tyrosine kinases Tie-1 and Tie-2 in blood vessel formation. Nature. 1995;**376**(6535):70-74

[80] Gustafsson T, Kraus WE. Exercise-induced angiogenesis-related growth and transcription factors in skeletal muscle, and their modification in muscle pathology. Frontiers in Bioscience. 2001;**6**:D75-D89

[81] Holash J et al. Vessel cooption, regression, and growth in tumors mediated by angiopoietins and VEGF. Science. 1999;**284**(5422):1994-1998

[82] Carmeli E et al. Matrix metalloproteinases and skeletal muscle: A brief review. Muscle & Nerve. 2004;**29**(2):191-197

[83] John A, Tuszynski G. The role of matrix metalloproteinases in tumor angiogenesis and tumor metastasis. Pathology Oncology Research. 2001;**7**(1):14

[84] Carmeli E et al. High intensity exercise increases expression of matrix metalloproteinases in fast skeletal muscle fibres. Experimental Physiology. 2005;**90**(4):613-619

[85] Brooks PC, Clark RA, Cheresh DA. Requirement of vascular integrin alpha v beta 3 for angiogenesis. Science. 1994;**264**(5158):569-571

[86] Brooks PC et al. Integrin alpha v beta 3 antagonists promote tumor regression by inducing apoptosis of angiogenic blood vessels. Cell. 1994;**79**(7):1157-1164

[87] Ouchi N et al. Follistatin-like 1, a secreted muscle protein, promotes endothelial cell function and revascularization in ischemic tissue through a nitric-oxide synthase-dependent mechanism. The Journal of Biological Chemistry. 2008;**283**(47):32802-32811

[88] Ernkvist M et al. Differential roles of p80- and p130-angiomotin in the switch between migration and stabilization of endothelial cells. Biochimica et Biophysica Acta. 2008;**1783**(3):429-437

[89] Roudier E et al. Angiomotin p80/p130 ratio: A new indicator of exercise-induced angiogenic activity in skeletal muscles from obese and non-obese rats? The Journal of Physiology. 2009;**587**(Pt 16):4105-4119

[90] Holland SJ et al. Bidirectional signalling through the EPH-family receptor Nuk and its transmembrane ligands. Nature. 1996;**383**(6602):722-725

[91] Wang HU, Chen ZF, Anderson DJ. Molecular distinction and angiogenic interaction between embryonic arteries and veins revealed by ephrin-B2 and its receptor Eph-B4. Cell. 1998;**93**(5):741-753

[92] O'Reilly MS et al. Endostatin: An endogenous inhibitor of angiogenesis and tumor growth. Cell. 1997;**88**(2):277-285

[93] Sudhakar A et al. Human tumstatin and human endostatin exhibit distinct antiangiogenic activities mediated by alpha v beta 3 and alpha 5 beta 1 integrins. Proceedings of the National Academy of Sciences of the United States of America. 2003;**100**(8):4766-4771

[94] Rege TA, Fears CY, Gladson CL. Endogenous inhibitors of angiogenesis in malignant gliomas: Nature's antiangiogenic therapy. Neuro-Oncology. 2005;**7**(2):106-121

[95] Malek MH, Olfert IM. Global deletion of thrombospondin-1 increases cardiac and skeletal muscle capillarity and exercise capacity in mice. Experimental Physiology. 2009;**94**(6):749-760

[96] Ribatti D et al. Human recombinant interferon alpha-2a inhibits angiogenesis of chick area vasculosa in shell-less culture. International Journal of Microcirculation, Clinical and Experimental. 1996;**16**(4):165-169

[97] Singh RK et al. Interferons alpha and beta down-regulate the expression of basic fibroblast growth factor in human carcinomas. Proceedings of the National Academy of Sciences of the United States of America. 1995;**92**(10):4562-4566

Chapter 2

Excitability of Vascular Smooth Muscle

Alexandra V. Ulyanova

Abstract

Regulation of pressure and local blood flow occurs at the level of resistance arteries and arterioles. Under physiological conditions, these small vessels exist in a state of partial constriction, termed myogenic tone. Myogenic tone is considered to be an intrinsic property of arteriolar smooth muscle cells, which membranes depolarize in response to increase in the intraluminal pressure. Oscillations of membrane potential in smooth muscles are mediated by the activity of voltage-gated L-type Ca^{2+} channels, which provide an influx of Ca^{2+} to activate various voltage-gated and Ca^{2+}-sensitive channels of smooth muscle cells and to initiate endothelial Ca^{2+} signaling needed for vasodilation. Although a relationship between change in membrane potential and myogenic response is considered to be universal throughout various smooth muscle tissues, it may be regulated differently based on autoregulatory responses and channels expression. Here we review electrophysiological signature of arteriolar smooth muscle in various tissues, with an emphases and specific examples of the excitability of 4th order arterioles isolated from skeletal muscle.

Keywords: vasculature, arteriolar smooth muscle, excitability, electrophysiology, voltage-gated Ca^{2+} and Na^{+} channels

1. Introduction

Regulation of pressure and local blood flow occurs at the level of resistance arteries and arterioles. Once blood exits the heart, it first flows into large elastic arteries, followed by smaller distributing arteries, which branch further into small resistance arteries and, finally, arterioles. The branching and reduction in vessel diameter actually results in the increase in total cross-sectional area of circulation. Because of their small diameter, resistance arteries and arterioles are the place of the largest pressure drop.

2. Myogenic tone

Resistance arteries and arterioles typically exhibit a state of partial constriction termed myogenic tone [1]. Myogenic tone is related to the level of the intraluminal pressure and provides a level of tone that vasodilators can act upon [2]. When the intraluminal pressure increases, resistance arteries and arterioles first dilate due to their elastic properties and then constrict to the new steady-state level. Myogenic tone is an intrinsic property of arteriolar smooth muscle cells and does not require endothelium [3, 4]. Resistance to blood flow is actively controlled by contraction or relaxation of arteriolar smooth muscle cells wrapped around the vessel so that their tone regulates the vessel diameter [2]. Arterial walls are made up of three layers: the tunica intima, tunica media and tunica adventitia. While the tunica intima contains endothelial cells and a thin layer of connective tissue, the tunica media supplies mechanical strength and contractile power. It is composed of several layers of spindle-shaped smooth muscle cells arranged helically in a matrix of elastin and collagen fibers. In some places, endothelial cells make contacts with smooth muscle cells to transmit signals between the intima and media. The tunica adventitia is mostly a connective tissue sheath with no distinct outer border. Its role is to tether the vessel loosely to the surrounding tissue [5]. Arterioles, the smallest resistance arteries placed right before capillaries, have a single layer of spindle-shaped smooth muscle cells [6]. Endothelial cells frame the lumen of arterioles. The shape and orientation within the vessels help to distinguish between these two distinct cells types (**Figure 1**).

Development of the tone is associated with depolarization of smooth muscle cells. While the mechanisms underlying depolarization are not completely understood, it is known that development of myogenic tone depends on extracellular Ca^{2+}. At 0 Ca^{2+}, pressurized resistance arteries and arterioles are fully dilated. Adding Ca^{2+} up to ≈ 2 mM to the extracellular space causes maximal tone [7]. In addition to the development of basal tone, the myogenic mechanism is believed to underlie the response to acute changes in pressure and contribute to spontaneous vasomotor activity [7].

Myogenic tone is controlled by intrinsic as well as extrinsic mechanisms. Intrinsic mechanisms include constriction of arterioles in response to pressure increase (Bayliss effect), endothelial secretions (nitric oxide, EDHF, prostacyclin, endothelin), vasoactive metabolites (e.g. adenosine in exercising muscle), autacoids (local vasoactive paracrine secretions such as histamine), and temperature. Important physiological responses mediated entirely by intrinsic regulation include the autoregulation of flow, and functional and reactive hyperemia. Intrinsic regulation also contributes to pathological responses such as inflammation and arterial vasospasm. Extrinsic regulation is brought about by factors originating outside the organ, namely the vasomotor nerves (sympathetic, parasympathetic and others) and circulating hormones such as adrenaline, vasopressin and insulin (for review, see [8]).

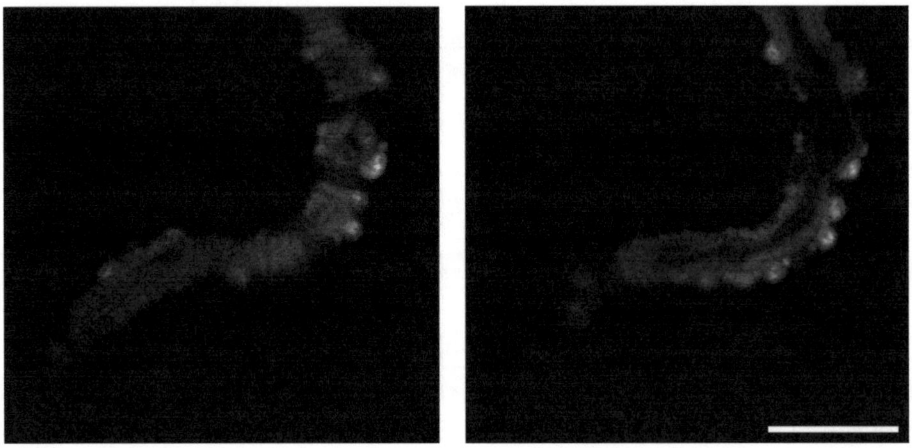

Figure 1.
4th order skeletal muscle arteriole loaded with 10 mM Fluo-4. A single layer of spindle-like vascular smooth muscle cells (VSMC) runs perpendicular to the vessel's length (left panel). Long endothelial cells (EC) frame the lumen of skeletal muscle arteriole (right panel, adapted from [6]). Scale bar = 50 μm.

3. Skeletal muscle vasculature

Skeletal muscle is the largest organ in the body that receives about 20% of cardiac output at rest and up to 80% during exercise. In skeletal muscle, the local blood flow is regulated over a 20× range to meet the demands of exercise. Therefore, vascular resistance of skeletal muscle is a critical determinant of total peripheral resistance and blood pressure. Arterioles of skeletal muscle have a high myogenic tone at rest as they are subject to major hyperemia (increased blood flow) during exercise [8]. In case of orthostatic hypotension, a potential contributor to cardiovascular adaptation to prolonged periods of bed rest or microgravity, skeletal muscle arterioles may develop functional or structural adaptations. Despite the physiological importance of skeletal muscle arterioles, little is known about ionic mechanisms underlying their excitability. Even within the same skeletal muscle, arterioles might respond differently to pressure changes. For instance, the first-order arterioles isolated from fast-twitch (e.g., white gastrocnemius) skeletal muscle fiber demonstrated both functional and structural changes such as reduced myogenic tone, decreased contractile responsiveness, and reduced wall thickness with no change in luminal diameter [9, 10]. In contrast, arterioles isolated from slow-twitch (e.g., soleus) fibers show no difference in myogenic tone or contractile responsiveness but rather a structural remodeling resulting in a decreased arteriole diameter [11].

4. Excitability of vascular smooth muscle

After development of isolated vessel techniques, the electrophysiology of smooth muscle cells was extensively studied in small arteries isolated from various vascular beds [12].

4.1 Membrane potential of vascular smooth muscle

Both slow changes (myogenic tone) and fast spikes in the membrane potential have been observed in arteriolar smooth muscle cells. Arteriolar smooth muscle cells can generate rhythmical contractions (vasomotion) over an extended period of time. Vasomotion occurs spontaneously or in response to vasoactive stimulation with a frequency of 3–20 per minute. The exact physiological role of vasomotion is not clear. In cases of ischemia and hypertension, it serves as a protective mechanism. Vasomotion is not a consequence of heartbeat, respiration, or neuronal input. It is generated within arteriolar smooth muscle cells by an endogenous pacemaker mechanism driven by a cytosolic Ca^{2+} oscillator. The cytosolic Ca^{2+} oscillator depends on Ca^{2+} entry and is regulated by transmitters and hormones, which increase the formation of $InsP_3$ and diacylglycerol (DAG) and promote oscillatory activity (for review, see [13]).

From the early studies, it has been appreciated that an increase in intraluminal pressure leads to depolarization and consequent contraction of vascular smooth muscle cell (electromechanical coupling). Mechanisms of depolarization are still not well understood. Depolarization increases Ca^{2+} concentration inside the cell, which leads to activation of myosin light chain kinase by Ca^{2+}/calmodulin and consequent contraction. Propagation of depolarization is achieved through gap junctions between neighboring smooth muscle cells. The signaling between endothelial and smooth muscle cells via gap junctions is essential for normal vascular function. Gap junctions in arteriolar smooth muscle cells and endothelial cells are formed by connexins Cx37, Cx40, and Cx43. Coupling between arteriolar smooth muscle cells appears to be essential for the maintenance of oscillations in membrane potential, the intracellular [Ca^{2+}], and vasomotion.

The exact signaling mechanisms that underlie detection of the mechanical stimulus and membrane depolarization are not completely understood (for review, see [7, 14]). Depolarization of vascular smooth muscle's membrane activates various voltage-gated ion channels, pumps and exchangers, including voltage-gated Ca^{2+} and K^+ channels, Ca^{2+}-activated BK_{Ca} and Cl_{Ca} channels, Na^+/Ca^{2+} exchanger, Ca^{2+}-ATP pump and N^+/K^+-ATPase, ATP-sensitive P2X receptor, and transient receptor potential (TRP) ion channels. Voltage-gated L-type Ca^{2+} channels are activated upon depolarization and increase Ca^{2+} concentration inside the cell. BK_{Ca} K^+ channels hyperpolarize smooth muscle cell back to its resting potential [15, 16]. Some vascular myocytes, particularly in large vessels, contract via pathways that appear to be unrelated to significant changes in the membrane potential (pharmacomechanical coupling).

Membrane potential of arteriolar smooth muscle cells is difficult to measure because of the changes in the intraluminal pressure and active vasomotor responses. Resting membrane potentials range from approximately −60 to −35 mV for physiological pressures (for review, see [17, 18]). Resting membrane potentials between −80 to −60 mV were previously reported for un-pressurized mice mesenteric artery [19], submucosal arterioles [20], and cerebral arterioles [21, 22]. A range of membrane potentials was reported for un-pressurized mice skeletal muscle arterioles (**Figure 2**). The average resting potential was reported to be around −68 mV (**Figure 2**, adapted from [6]). However, it's significantly more negative than −55 mV previously reported for un-pressurized arterioles in rat cremaster muscle [23]. Resting membrane potential in vascular smooth muscle cells is determined to a large extent by K^+ conductance [18]. Intracellular $[Cl^-]$ in vascular smooth muscle cells is around 55 mM, which is unusually high and mediated probably through Cl^-—HCO_3^- exchanger. A small influx of Ca^{2+} and efflux of Cl^- ions at rest reduce the membrane potential from the Nernst equilibrium for K^+ [24].

4.2 Excitability of vascular smooth muscle

Most of the arteriolar smooth muscle cells are quiescent. In some cases, a drop in the intraluminal pressure (during trauma to a vessel) that leads to hyperpolarization and dilation generates membrane action potentials. Physiological role of membrane action potentials in arteriolar smooth muscle is unclear, as relatively small changes in the membrane potential (between −55 and − 35 mV) are sufficient to control Ca^{2+} entry and to initiate contraction in response to stimulation by mechanical stress of the blood flow [14]. Spontaneous action potentials could be associated either with rhythmic vasomotor activity [23]; follow K^+-induced hyperpolarization and dilation [25], or precede injury-induced constriction [26].

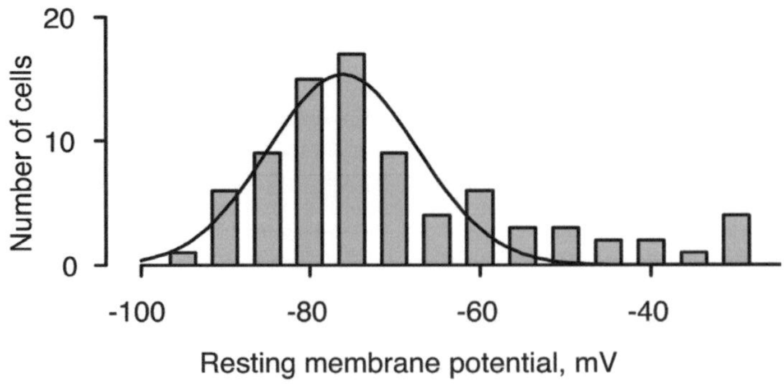

Figure 2.
Resting membrane potential of un-pressurized 4th order arterioles isolated from mice skeletal muscle. Distribution of the resting membrane potential values was fitted by a single Gaussian function peaking at −77 ± 2 mV (n = 81, smooth line). The average resting potential was V_{rest} *= −68 ± 2 mV (n = 81). Adapted from [6].*

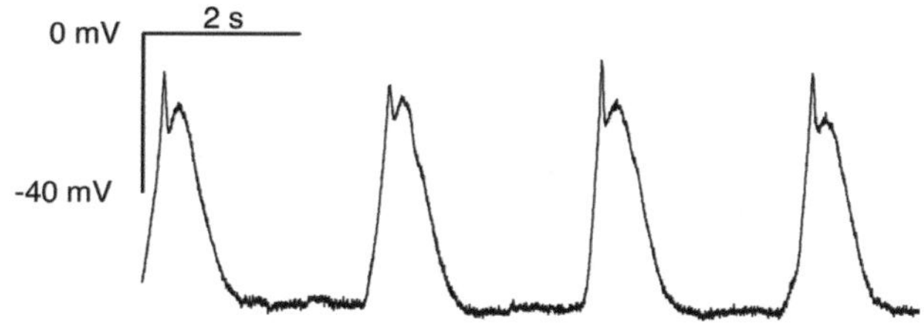

Figure 3.
Spontaneous action potentials were observed in un-pressurized skeletal muscle arterioles (2 mM Ca bath solution). Smooth muscle cells were current-clamped using gramicidin-perforated configuration, with intracellular solution containing 150 KCl. Adapted from [6].

Only a few recordings of action potentials in skeletal muscle arterioles were made so far [6, 23, 25, 27]. Spontaneous action potential spikes could be observed in pressurized small arteries [23, 25], as well as in un-pressurized arterioles as shown in **Figure 3** [6, 21].

4.3 Voltage-gated channels of arterial smooth muscle

Unlike in many other excitable tissues, action potentials in smooth muscle cells of arteries [28] and arterioles [21] are thought to be independent of voltage-gated Na^+ channels, as they could not be blocked by the application of voltage-gated Na^+ channel blocker tetrodotoxin (TTX). Depolarization of smooth muscle cells induced by the change in intraluminal pressure and/or tissue-stretch produces an increase in the intracellular [Ca^{2+}], consequent myosin light chain phosphorylation and contraction (for review, see [14]). Since specific blockers of L-type voltage-gated Ca^{2+} channels suppress both the upstroke and the after-depolarizing components of action potentials, these channels are thought to be the main pathway for the depolarizing current. Nevertheless, several groups have demonstrated that voltage-gated Na^+ channels are present in arterial beds and are activated upon membrane depolarization [6, 29].

4.3.1 Voltage-gated Ca^{2+} channels

Voltage-gated Ca^{2+} channels mediate influx of Ca^{2+} ions in response to membrane depolarization and regulate intracellular processes such as contraction, secretion, neurotransmission, and gene expression in many different cell types. Their activity is essential to couple electrical signals in the cell surface to physiological events in cells. Voltage-gated Ca^{2+} channels are comprised of the pore-forming α_1 subunit in complex with auxiliary subunits. A transmembrane disulfide-linked $\alpha_2\delta$, an intracellular β, and a γ subunit are components of most types of calcium channels. The α_1 subunit contains four repeats of a domain with six transmembrane segments, the fourth of which is the voltage-sensing S4 segment. The pore loop between transmembrane segments S5 and S6 in each domain determines ion conductance and selectivity (for review, see [30]). Ten distinct genes encode mammalian α_1 subunits of three subfamilies of voltage-gated Ca^{2+} channels. The amino acid sequences of these α_1 subunits are more than 70% identical within a subfamily but less than 40% identical among subfamilies. Voltage-gated Ca^{2+} channels are named using the chemical symbol of the principal permeating ion with the principal physiological regulator (voltage) indicated as a subscript according to the nomenclature developed for voltage-gated K^+ channels [31]. The Ca_V1 subfamily ($Ca_V1.1$–$Ca_V1.4$) represents high-voltage activated L-type Ca^{2+} channels, the Ca_V2 subfamily ($Ca_V2.1$–$Ca_V2.3$) represents neuronal N-, P/Q-, and R-types Ca^{2+} channels, and Ca_V3 subfamily ($Ca_V3.1$–$Ca_V3.3$) represents low-voltage activated T-type Ca^{2+} channels.

The pharmacology and biophysics of Ca^{2+} channels subfamilies are quite distinct. L-type Ca^{2+} calcium channels typically require a strong depolarization for activation and do not inactivate at positive potentials. They are the main pathway for Ca^{2+} currents recorded in muscle cells, where they initiate contraction and secretion. The Ca_V1 subfamily is the molecular target of the organic Ca^{2+} channel blockers used widely in the therapy of cardiovascular diseases. L-type Ca^{2+} channels are blocked by the organic antagonists, including dihydropyridines (e.g., nifedipine), phenylalkylamines, and benzothiazepines. Dihydropyridines can be channel activators or inhibitors and therefore are thought to act allosterically to shift the channel toward the open or closed state rather than by occluding the pore. T-type Ca^{2+} channels are activated by weak depolarization and are transient at sustained depolarization. They are expressed in a wide variety of cell types, where they are involved in shaping the action potential and controlling patterns of repetitive firing. The Ca_V3 subfamily of voltage-gated Ca^{2+} channels is insensitive to both the dihydropyridines that block Ca_V1 channels and the spider and cone snail toxins that block the Ca_V2 channels. There are no widely useful pharmacological agents that block T-type Ca^{2+} currents specifically. Mibefradil blocks both T-type Ca^{2+} channels and with less potency L-type Ca^{2+} channels. Currents through expressed $Ca_V3.2$ channels could be selectively blocked by application of 40 μM of Ni^{2+} [32].

Currents through dihydropyridine-sensitive Ca^{2+} channels were recorded in arteries of various vascular beds [21, 33–36], including from smooth muscle cells of skeletal muscle arterioles (**Figure 4**, left panel). $Ca_V1.2$ considered to be the principal sub-type of voltage-gated Ca^{2+} channels involved in excitation-contraction coupling of vascular smooth muscle cells [37]. Since L-type Ca^{2+} blocker nifedipine did not eliminate basal tone in skeletal muscle arterioles, other Ca^{2+} entry mechanisms are believed to contribute to myogenic tone along with L-type Ca^{2+} channels [38]. In addition to L-type, vascular smooth muscle cells also express T-type Ca^{2+} channels (for review, see [30, 39]). Currents through T-type Ca^{2+} channels have been found in smooth muscle cells of arteries and arterioles of cerebral [21], mesenteric [40], renal [41], coronary [35], and skeletal [6] vascular beds as shown in **Figure 4** (right panel). Although the messenger RNAs for both, $Ca_V3.1$ and $Ca_V3.2$ T-type Ca^{2+} channels were found in rat cremaster arterioles [42], only L-type Ca^{2+} currents were recorded in single smooth muscle cells isolated from resistance arteries of hamster cremaster muscle [43]. The pressure/stretch stimulus initiates a depolarization that

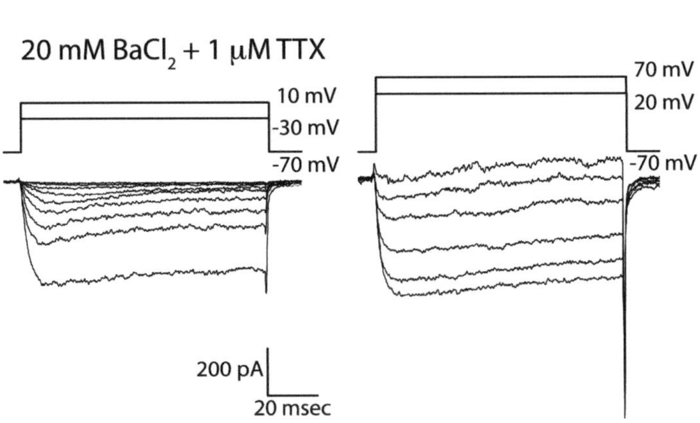

Figure 4.
Voltage-gated T-type and L-type Ca2+ channels are present in arteriolar smooth muscle. Whole-cell Ba^{2+} currents were recorded in the presence of 1 μM tetrodotoxin (TTX), voltage-gated Na^+ channel blocker as described in more detail in [6]. Two kinetically different components were observed: while voltage steps from −30 up to −10 mV produced fast inactivating T-type Ca^{2+} current (left panel), further depolarization activated slowly inactivating L-type Ca^{2+} current (right panel). Adapted from [6].

causes Ca^{2+} influx through voltage-gated L-type Ca^{2+} channels and initiates Ca^{2+} sparklets [44–46]. Both, Ca^{2+} sparklets and Ca^{2+} sparks (Ca^{2+} release from intracellular stores) signaling events, activate negative feedback mechanisms via Ca^{2+}-dependent K^+ currents thus preventing unrestrained depolarization [47–49]. While Ca^{2+} influx through L-type Ca^{2+} channels are believed to be involved in contraction of vascular smooth muscle, T-type Ca^{2+} channels, particularly $Ca_V3.2$ appear to be mostly involved in relaxation of coronary arteries, acting through activation of BK_{Ca} channels [40–43, 50, 51].

4.3.2 Voltage-gated Na^+ channels

In addition to voltage-gated Ca^{2+} channels, significant TTX-sensitive Na^+ currents were found in smooth muscle cells from rabbit main pulmonary artery [52], human aorta [53], murine mesenteric arteries [29], and skeletal muscle arterioles [6] as shown in **Figure 5**. The above observations suggest that more complex mechanisms are likely to generate action potential in vascular smooth muscle cells. When voltage-gated Ca^{2+} and Na^+ channels coexist in the same cell, Na^+ conductance is thought to be responsible for generation and propagation of action potential in excitable cells such as neurons, striated muscle, and neuroendocrine cells. It has lower threshold and faster rise time than Ca^{2+} conductance. After Na^+ channels rapidly inactivate, L-type voltage-gated Ca^{2+} channels open and further depolarize the cell, thus prolonging the plateau of an action potential. In contrast, T-type voltage-gated Ca^{2+} channels may open at resting potential and produce depolarizing current that brings the cell to threshold for Na^+ spike [54, 55].

Voltage-gated Na^+ channels consist of the subunit associated with auxiliary subunits. The pore-forming subunit is sufficient for functional expression, but the kinetics and voltage dependence of channel gating are modified by the subunits. These auxiliary subunits are involved in channel localization and interaction with cell adhesion molecules, extracellular matrix, and intracellular cytoskeleton. Similar to the voltage-gated Ca^{2+} channels, the subunits of Na^+ channels are organized in four homologous domains, each composed of six transmembrane segments. S4 segment is a voltage sensor, and a pore loop located between the S5 and S6 segments determines ion conductance and selectivity (for review, see [56]).

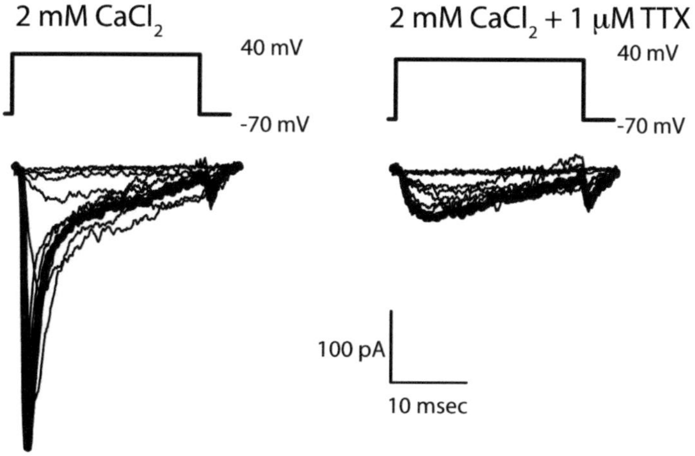

Figure 5.
TTX-sensitive voltage-gated Na^+ channels are present in arteriolar smooth muscle. In arteriolar smooth muscle of skeletal muscle arterioles, voltage steps produced inward currents with at least two kinetically distinct components in 2 mM Ca solution (left panel). The fast component was through voltage-gated Na^+ channels as it was blocked by application of 1 μM TTX (right panel). Adapted from [6].

According to the nomenclature, the Na_V1 superfamily ($Na_V1.1$–$Na_V1.9$) represents voltage-gated Na^+ channels. Unlike the different classes of voltage-gated Ca^{2+} channels, the functional properties of Na^+ channels are relatively similar. Amino acid sequences of the nine mammalian Na^+ channel isoforms are more than 50% identical to each other, but separate families are difficult to define. By this criterion, nine isoforms are considered to be members of one Na_V1 gene subfamily. Some voltage-gated Na^+ channels ($Na_V1.1$, $Na_V1.2$, $Na_V1.3$, $Na_V1.4$, $Na_V1.6$, and $Na_V1.7$) could be blocked by tetrodotoxin that binds to the extracellular side of the pore.

Several isoforms of TTX-sensitive voltage-gated channels have been found previously smooth muscle cells of in vasa recta ($Na_V1.3$), portal vein ($Na_V1.6$ and $Na_V1.8$), and vas deference ($Na_V1.6$) [57–59]. The activity of these voltage-gated Na^+ channels may be tightly controlled by elevations of intracellular Ca^{2+}, potentially suppressing activity of Na^+ channels [6]. For instance, Ca^{2+}/CaM binding was shown to down-regulates skeletal muscle isoform $Na_V1.4$ by shifting its steady-state inactivation curve in the hyperpolarizing direction [60]. $Na_V1.3$ channels in descending vasa recta are suppressed by calmodulin inhibitors, while elevation of the intracellular [Ca^{2+}] shifts the voltage-dependence of their activation to more positive voltages [57]. In addition, Ca^{2+}-dependent down-regulation of Na^+ channels can also occur via the PKC pathway since activation of PKC decreases peak sodium currents through brain $Na_V1.2$ and skeletal muscle $Na_V1.4$ channels by up to 80% [61, 62].

5. Conclusion

L-type voltage-gated Ca^{2+} channels are involved in excitation-contraction coupling of vascular smooth muscle cells. In addition, T-type Ca^{2+} channels were detected in arteriolar smooth muscle cells. These two types of voltage-gated Ca^{2+} channels together play an important role in the constriction/relaxation of smooth muscle cells by regulating Ca^{2+} signaling during myogenic tone. However, there are indications that other type of voltage-gated channels, specifically Na^+ channels are present in various vascular beds. While the role of voltage-gated Ca^{2+} channels is well established, contribution of voltage-gated Na^+ channel remains to be determined.

Acknowledgements

I would like to thank Dr. Roman E. Shirokov for invaluable discussion related to this work.

Conflict of interest

The author declares no conflict of interest.

Abbreviations

DAG	diacylglycerol
EC	endothelial cell
EDHF	endothelium-derived hyperpolarizing factor
$InsP_3$	inositol 1,4,5-triphosphate
TTX	tetrodotoxin
VSMC	vascular smooth muscle cell

Author details

Alexandra V. Ulyanova
Center for Brain Injury and Repair, Department of Neurosurgery, Perelman School of Medicine, University of Pennsylvania, Philadelphia, USA

*Address all correspondence to: ulyanova@pennmedicine.upenn.edu.

References

[1] Bayliss N. On the local reactions of the arterial wall to changes of internal pressure. The Journal of Physiology. 1902;**28**:220-231

[2] Mellander S, Johansson B. Control of resistance, exchange, and capacitance functions in the peropheral circulation. Pharmacological Reviews. 1968;**20**(3):117-196

[3] Kuo L, Chilian WM, Davis MJ. Coronary arteriolar myogenic response is independent of endothelium. Circulation Research. 1990;**66**(3):860-866

[4] Falcone JC, Davis MJ, Meininger GA. Endothelial independence of myogenic response in isolated skeletal muscle arterioles. American Journal of Physiology. Heart and Circulatory Physiology. 1991;**260**(1):H130-H135

[5] Greensmith JE, Duling BR. Morphology of the constricted arteriolar wall: Physiological implications. American Journal of Physiology. Heart and Circulatory Physiology. 1984;**247**(5):H687-HH98

[6] Ulyanova AV, Shirokov RE. Voltage-dependent inward currents in smooth muscle cells of skeletal muscle arterioles. PLoS One. 2018;**13**(4):e0194980

[7] Davis MJ, Hill MA. Signaling mechanisms underlying the vascular myogenic response. Physiological Reviews. 1999;**79**(2):387-423

[8] Herring N, Paterson DJ. Levick's Introduction to Cardiovascular Physiology. 6th Ed. Boca Raton, FL: CRC Press; 2018

[9] Delp MD, Colleran PN, Wilkerson MK, McCurdy MR, Muller-Delp J. Structural and functional remodeling of skeletal muscle microvasculature is induced by simulated microgravity. American Journal of Physiology. Heart and Circulatory Physiology. 2000;**278**(6):H1866-H1873

[10] Delp MD. Myogenic and vasoconstrictor responsiveness of skeletal muscle arterioles is diminished by hindlimb unloading. Journal of Applied Physiology. 1999;**86**(4):1178-1184

[11] Heaps CL, Bowles DK. Nonuniform changes in arteriolar myogenic tone within skeletal muscle following hindlimb unweighting. Journal of Applied Physiology. 2002;**92**(3):1145-1151

[12] Duling BR, Gore RW, Dacey RG Jr, Damon DN. Methods for isolation, cannulation, and in vitro study of single microvessels. American Journal of Physiology. Heart and Circulatory Physiology. 1981;**241**(1):H108-H116

[13] Berridge MJ. Smooth muscle cell calcium activation mechanisms. The Journal of Physiology. 2008;**586**(21):5047-5061

[14] Hill MA, Davis MJ, Meininger GA, Potocnik SJ, Murphy TV. Arteriolar myogenic signalling mechanisms: Implications for local vascular function. Clinical Hemorheology and Microcirculation. 2006;**34**(1):67-79

[15] Brayden J, Nelson M. Regulation of arterial tone by activation of calcium-dependent potassium channels. Science. 1992;**256**(5056):532-535

[16] Jaggar JH, Porter VA, Lederer WJ, Nelson MT. Calcium sparks in smooth muscle. American Journal of Physiology. Cell Physiology. 2000;**278**(2):C235-C256

[17] Hirst GD, Edwards FR. Sympathetic neuroeffector transmission in arteries and arterioles. Physiological Reviews. 1989;**69**(2):546-604

[18] Nelson MT, Patlak JB, Worley JF, Standen NB. Calcium channels, potassium channels, and voltage dependence of arterial smooth muscle tone. American Journal of Physiology. Cell Physiology. 1990;**259**(1):C3-C18

[19] Kuriyama H, Suzuki H. Adrenergic transmissions in the guinea-pig mesenteric artery and their cholinergic modulations. The Journal of Physiology. 1981;**317**:383-396

[20] Hirst GD, van Helden DF. Ionic basis of the resting potential of submucosal arterioles in the ileum of the guinea-pig. The Journal of Physiology. 1982;**333**:53-67

[21] Hirst GD, Silverberg GD, van Helden DF. The action potential and underlying ionic currents in proximal rat middle cerebral arterioles. The Journal of Physiology. 1986;**371**:289-304

[22] Knot HJ, Nelson MT. Regulation of arterial diameter and wall [Ca^{2+}] in cerebral arteries of rat by membrane potential and intravascular pressure. The Journal of Physiology. 1998;**508**(Pt 1):199-209

[23] Kotecha N, Hill MA. Myogenic contraction in rat skeletal muscle arterioles: Smooth muscle membrane potential and Ca^{2+} signaling. American Journal of Physiology. Heart and Circulatory Physiology. 2005;**289**(4):H1326

[24] Gerstheimer FP, Mühleisen M, Nehring D, Kreye VA. A chloride-bicarbonate exchanging anion carrier in vascular smooth muscle of the rabbit. Pflügers Archiv. 1987;**409**(1-2):60-66

[25] Burns WR, Cohen KD, Jackson WF. K^+-induced dilation of hamster cremasteric arterioles involves both the Na^+/K^+-ATPase and inward-rectifier K^+ channels. Microcirculation. 2004;**11**(3):279-293

[26] Graham JM, Keatinge WR. Responses of inner and outer muscle of the sheep carotid artery to injury. The Journal of Physiology. 1975;**247**(2):473-482

[27] Bartlett IS, Crane GJ, Neild TO, Segal SS. Electrophysiological basis of arteriolar vasomotion in vivo. Journal of Vascular Research. 2000;**37**(6):568-575

[28] Keatinge WR. Sodium flux and electrical activity of arterial smooth muscle. The Journal of Physiology. 1968;**194**(1):183-200

[29] Berra-Romani R, Blaustein MP, Matteson DR. TTX-sensitive voltage-gated Na^+ channels are expressed in mesenteric artery smooth muscle cells. American Journal of Physiology. Heart and Circulatory Physiology. 2005;**289**(1):H137-H145

[30] Catterall WA, Perez-Reyes E, Snutch TP, Striessnig J. International union of pharmacology. XLVIII. Nomenclature and structure-function relationships of voltage-gated calcium channels. Pharmacological Reviews. 2005;**57**(4):411-425

[31] Chandy KG, Gutman GA. Nomenclature for mammalian potassium channel genes. Trends in Pharmacological Sciences. 1993;**14**(12):434

[32] Lee J-H, Gomora JC, Cribbs LL, Perez-Reyes E. Nickel block of three cloned T-type calcium channels: Low concentrations selectively block alpha1H. Biophysical Journal. 1999;**77**(6):3034-3042

[33] Aaroson PI, Bolton TB, Lang RJ, MacKenzie I. Calcium currents in single isolated smooth muscle cells from the rabbit ear artery in normal-calcium and high-barium solutions. The Journal of Physiology. 1988;**405**(1):57-75

[34] Bean B, Sturek M, Puga A, Hermsmeyer K. Calcium channels in muscle cells isolated from rat mesenteric

arteries: Modulation by dihydropyridine drugs. Circulation Research. 1986;**59**(2):229-235

[35] Ganitkevich VY, Isenberg G. Contribution of two types of calcium channels to membrane conductance of single myocytes from guinea-pig coronary artery. The Journal of Physiology. 1990;**426**(1):19-42

[36] Smirnov SV, Aaronson PI. Ca^{2+} currents in single myocytes from human mesenteric arteries: Evidence for a physiological role of L-type channels. The Journal of Physiology. 1992;**457**(1):455-475

[37] Koch W, Ellinor P, Schwartz A. cDNA cloning of a dihydropyridine-sensitive calcium channel from rat aorta. Evidence for the existence of alternatively spliced forms. The Journal of Biological Chemistry. 1990;**265**(29):17786-17791

[38] Hill MA, Meininger GA. Calcium entry and myogenic phenomena in skeletal muscle arterioles. American Journal of Physiology. Heart and Circulatory Physiology. 1994;**267**(3):H1085-H1H92

[39] Perez-Reyes E. Molecular physiology of low voltage-activated T-type calcium channels. Physiological Reviews. 2003;**83**(1):117-161

[40] Morita H, Cousins H, Onoue H, Ito Y, Inoue R. Predominant distribution of nifedipine-insensitive, high voltage-activated Ca^{2+} channels in the terminal mesenteric artery of guinea pig. Circulation Research. 1999;**85**(7):596-605

[41] Gordienko DV, Clausen C, Goligorsky MS. Ionic currents and endothelin signaling in smooth muscle cells from rat renal resistance arteries. American Journal of Physiology. Renal Physiology. 1994;**266**(2):F325-F341

[42] VanBavel E, Sorop O, Andreasen D, Pfaffendorf M, Jensen BL. Role of T-type calcium channels in myogenic tone of skeletal muscle resistance arteries. American Journal of Physiology. Heart and Circulatory Physiology. 2002;**283**(6):H2239-H2H43

[43] Cohen KD, Jackson WF. Hypoxia inhibits contraction but not calcium channel currents or changes in intracellular calcium in arteriolar muscle cells. Microcirculation. 2003;**10**(2):133-141

[44] Amberg GC, Navedo MF, Nieves-Cintron M, Molkentin JD, Santana LF. Calcium sparklets regulate local and global calcium in murine arterial smooth muscle. Journal of Physiology (London). 2007;**579**(1):187-201

[45] Navedo MF, Amberg GC, Nieves M, Molkentin JD, Santana LF. Mechanisms underlying heterogeneous Ca^{2+} sparklet activity in arterial smooth muscle. The Journal of General Physiology. 2006;**127**(6):611-622

[46] Navedo MF, Amberg GC, Votaw VS, Santana LF. Constitutively active L-type Ca^{2+} channels. PNAS. 2005;**102**(31):11112-11117

[47] Jaggar JH, Wellman GC, Heppner TJ, Porter VA, Perez GJ, Gollasch M, et al. Ca^{2+} channels, ryanodine receptors and Ca^{2+}-activated K^{+} channels: A functional unit for regulating arterial tone. Acta Physiologica Scandinavica. 2008;**164**(4):577-587

[48] Ledoux J, Werner ME, Brayden JE, Nelson MT. Calcium-activated potassium channels and the regulation of vascular tone. Physiology. 2006;**21**(1):69-78

[49] Nelson MT, Cheng H, Rubart M, Santana LF, Bonev AD, Knot HJ, et al. Relaxation of arterial smooth muscle by calcium sparks. Science. 1995;**270**(5236):633

[50] Chen CC, Lamping KG, Nuno DW, Barresi R, Prouty SJ, Lavoie JL, et al.

Abnormal coronary function in mice deficient in alpha1H T-type Ca^{2+} channels. Science. 2003;**302**(5649):1416-1418

[51] Morita H, Shi J, Ito Y, Inoue R. T-channel-like pharmacological properties of high voltage-activated, nifedipine-insensitive Ca^{2+} currents in the rat terminal mesenteric artery. British Journal of Pharmacology. 2002;**137**(4):467-476

[52] Okabe K, Kitamura K, Kuriyama H. The existence of a highly tetrodotoxin sensitive Na channel in freshly dispersed smooth muscle cells of the rabbit main pulmonary artery. Pflügers Archiv—European Journal of Physiology. 1988;**411**(4):423-428

[53] Cox RH, Zhou Z, Tulenko TN. Voltage-gated sodium channels in human aortic smooth muscle cells. Journal of Vascular Research. 1998;**35**(5):310-317

[54] Llinás R, Yarom Y. Electrophysiology of mammalian inferior olivary neurones in vitro. Different types of voltage-dependent ionic conductances. The Journal of Physiology. 1981;**315**:549-567

[55] Llinás R, Yarom Y. Properties and distribution of ionic conductances generating electroresponsiveness of mammalian inferior olivary neurones in vitro. The Journal of Physiology. 1981;**315**:569-584

[56] Catterall WA, Goldin AL, Waxman SG. International union of pharmacology. XLVII. Nomenclature and structure-function relationships of voltage-gated sodium channels. Pharmacological Reviews. 2005;**57**(4):397

[57] Lee-Kwon W, Goo JH, Zhang Z, Silldorff EP, Pallone TL. Vasa recta voltage-gated Na^+ channel $Na_V1.3$ is regulated by calmodulin. American Journal of Physiology. Renal Physiology. 2007;**292**(1):F404-F414

[58] Saleh S, Yeung SYM, Prestwich S, Pucovsky V, Greenwood I. Electrophysiological and molecular identification of voltage-gated sodium channels in murine vascular myocytes. The Journal of Physiology. 2005;**568**(1):155-169

[59] Zhu H-L, Shibata A, Inai T, Nomura M, Shibata Y, Brock JA, et al. Characterization of $Na_V1.6$-mediated Na^+ currents in smooth muscle cells isolated from mouse vas deferens. Journal of Cellular Physiology. 2010;**223**(1):234-243

[60] Biswas S, Deschenes I, DiSilvestre D, Tian Y, Halperin VL, Tomaselli GF. Calmodulin regulation of $Na_V1.4$ current: Role of binding to the carboxyl terminus. The Journal of General Physiology. 2008;**131**(3):197-209

[61] Numann R, Catterall WA, Scheuer T. Functional modulation of brain sodium channels by protein kinase C phosphorylation. Science. 1991;**254**(5028):115-118

[62] Numann R, Hauschka S, Catterall W, Scheuer T. Modulation of skeletal muscle sodium channels in a satellite cell line by protein kinase C. The Journal of Neuroscience. 1994;**14**(7):4226-4236

Chapter 3

Orexin System and Avian Muscle Mitochondria

Kentu Lassiter and Sami Dridi

Abstract

In mammals, orexin A and B (also known as hypocretin 1 and 2) are two orexigenic peptides produced primarily by the lateral hypothalamus that signal through two G-protein-coupled receptors, orexin receptors 1/2, and have been implicated in the regulation of several physiological processes. However, the physiological roles of orexin are not well defined in avian (non-mammalian vertebrate) species. Recently, we made a breakthrough by identifying that orexin and its related receptors 1/2 (ORXR1/2) are expressed in avian muscle tissue and cell line, and appears to be a secretory protein. Functional in vitro studies showed that orexin A and B differentially regulated expression of the orexin system, suggesting that orexins might have autocrine, paracrine, and/or endocrine roles. Administration of recombinant orexin modulated mitochondrial biogenesis, dynamics, function, and bioenergetics. In this chapter, we include a brief overview of the (patho) physiological role of orexin, comparative findings between mammalian and avian orexin, and in-depth analysis of orexin's action on avian muscle mitochondria.

Keywords: orexin system, muscle, avian species, mitochondria, gene expression

1. Introduction

Orexin/hypocretin was originally identified by two different groups and reported in 1998 as an orexigenic feeding-related neuropeptide mainly produced in the rat hypothalamus [1, 2]. Numerous subsequent studies conducted in mammals have shown that orexin and its receptors regulate several physiological processes including food and water intake [3], control of wakefulness [4], circadian clock [5], energy and glucose homeostasis [6–8], lipid metabolism [9, 10], heart rate and blood pressure [11, 12], and neuroendocrine response to stress [13]. Despite these advancements in understanding the orexin system, studies investigating its distribution and function in avian (non-mammalian vertebrate) species are very limited and merit more consideration and further in-depth explorations. Understanding orexin distribution and unraveling its function in avian tissues, particularly in broiler (meat-type chickens) skeletal muscle is of particular importance not only to the poultry industry, but also to the biomedical field.

The poultry (meat and egg) industry supports the livelihoods and food security of billions of people worldwide with an average annual production of 118 million metric tons of meat and 1360 billion eggs in 2017 [14]. This success was mainly achieved by intensive genetic selection for important economic and agricultural traits such as high growth rate, improved feed efficiency (conversion of feed to muscle mass), and increased muscle mass [15]. Today, indeed, modern broilers are marketed in about half the time and at about twice the body weight compared to 50–60 years ago [16]. This fast growth is largely allocated to the beast muscle [17].

Chickens are naturally hyperglycemic compared with mammals [18], insulin-resistant [19], lack the glucose transporter GLUT4 [20], lack functional brown adipose tissue [21], and are prone to obesity [22], and thereby they represent a highly relevant animal model for biomedical researches.

The proper function of mitochondria is invariably related to muscle growth since these organelles produce a vast majority (~90%) of the ATP needed for tissue growth and maintenance of energy metabolism homeostasis. Previous studies in broiler chickens have shown that mitochondrial perturbations are directly related to a decrease in feed efficiency. Additionally, the absence of orexin's effect on feed intake in chickens [23] suggest that this neuropeptide may be more involved in other peripheral metabolic and physiological processes rather than food/water intake and wakefulness that is seen in mammals. Therefore, if orexin and its receptors are shown to be present in skeletal muscle tissue of avian species, they may be exerting control of energy metabolism and muscle growth (i.e., myogenesis, insulin sensitivity, and glucose transport) in the cell in part by regulating mitochondrial dynamics. The role of orexin in skeletal muscle function could be of interest not only in improving health and feed efficiency of agricultural animals, but also in molecular medicine for pathophysiological understanding and therapeutic perspectives.

2. Identification of orexin and its receptors

Orexin, which regulates wakefulness, energy homeostasis, and appetite/feeding behavior based on nutritional status, is a neuropeptide that was originally discovered in the hypothalamus of rats by two different research groups in 1998. De Lecea's group isolated cDNAs selectively expressed within the hypothalamus and found that two peptides showed high amino acid sequence homology with secretin (the gut peptide hormone), therefore they named the two peptides hypocretin 1 and 2 [1]. Sakurai's group, on the other hand, used reverse pharmacology to identify ligands of orphan G-protein-coupled receptors. Orphan receptors are those whose ligand and physiological actions are unknown [24]. They identified a novel family of neuropeptides that induced feeding behavior, so they named them orexin A and B [2]. The term orexin originates from the Greek word "orexis," meaning appetite. Both orexin A and B are formed by proteolytic cleavage of the precursor prepro-orexin [2]. When initially discovered in rats, the precursor peptide prepro-orexin was shown to be a 130-residue polypeptide from which the mature peptides ORX-A and ORX-B were formed, with ORX-A containing 33 amino acids and a molecular weight 3.562 kDa and ORX-B containing 28 amino acids and a molecular weight of 2.937 kDa [2]. When comparing the two peptides, ORX-B was shown to be 56% identical in amino acid sequence to ORX-A. However, when comparing mammalian species (human, rat, mouse, pig, and cow), the sequence and structure of both peptides is highly conserved [2]. A number of studies have also shown that the structures of ORX-A and ORX-B in chicken and certain types of fish are conserved when compared to their mammalian counterparts [25–28].

ORX-A and ORX-B signal through the G-protein-coupled receptors orexin receptor 1 (ORXR1) and orexin receptor 2 (ORXR2). These two ubiquitously expressed receptors were first identified in human brain tissue through expressed sequence tags combined with database searching using tBLASTn [2, 29]. In humans it has been shown that the amino acid sequence for ORXR1 and ORXR2 is more than

60% identical, making them more similar to each other than to other G-protein-coupled receptors [2]. The same study also showed that both receptors are highly conserved between humans and rats, with the sequence identity being greater than 90% for both. The two orexin peptides have different binding affinities for the two orexin receptors. ORX-A is able to bind to both receptors but has a higher affinity for ORXR1, while ORX-B binds to ORXR2 with the same affinity as ORX-A [28]. Several studies have indicated that the binding of orexins to orexin receptors activates multiple G-proteins. In studies conducted using humans [30, 31] and rats [32], it was shown that the binding of ORXR2 activates Gi, Gs, Go, and Gq proteins in adrenal cortical tissue. It appears that the responses to orexin receptor signaling are highly diverse. The activation of the various G-proteins can lead to a variety of cellular responses such as the regulation of protein/lipid kinases [33, 34]. In the case of orexin stimulation, activation of G-proteins can lead to the excitation of neurons that affect the regulation of ion channels, the activation of signaling cascades that regulate the activity of adenylyl cyclase and phospholipases, and activation of cell death pathways [35, 36].

3. Orexin system in avian species

Significantly fewer studies concerning the orexin system have been conducted in avian species when compared to mammals. Chicken prepro-orexin was first cloned, sequenced, and characterized in 2002 [37]. In that study, chicken orexin cDNA was shown to be expressed in the periventricular and lateral hypothalamic areas and consisting of 658 bp that encode 148 amino acids. Also, chicken ORX-A and ORX-B are evolutionary conserved with their mammalian counterparts, showing ~85 and 65% similarity at the amino acid level [37]. Characterization of the chicken orexin receptor shows that its cDNA has a length of 1869 bp that encode 501 amino acids, which corresponds to mammalian ORXR2 with an 80% homology [37]. Studies looking at tissue distribution of orexin and orexin receptors in chickens show that the peptides are expressed in the brain [37–40], pituitary gland, adrenal gland, testis and ovary [38], and the stomach and intestine [41].

Orexin does not appear to elicit the same responses in birds as it does in mammals. One of the most noted actions that centrally administered orexin has in mammals is that it stimulates feeding/food intake [2, 3, 28, 42]. However, central administration of ORX-A or ORX-B did not stimulate feed intake in neonatal broiler and layer chicks [23, 43] or adult pigeons [44]. Studies examining mRNA expression of prepro-orexin in the hypothalamus of chicken [37] and quail [45] following 24 h fasting showed no increase in expression, providing further evidence for the lack of a stimulatory effect on feeding behavior in birds. Another feeding behavior study did show an increase in prepro-orexin mRNA [46], but this was measured after 48 h fasting, which would be an extreme fasting condition for broiler chickens.

As stated previously, another hallmark of orexin function in mammals is its effects on the regulation of sleep/wakefulness, where a dysfunction in the orexin system is associated with the sleep condition narcolepsy [4]. Studies investigating the effects of orexin on arousal in birds have been conducted with mixed results. It has been concluded that either hypothalamic orexin does not play a role in arousal of the sleep/wake cycle [39], or that only ORX-A in conjunction with the enzyme monoamine oxidase-A (MAO-A) increases arousal in layer chicks only and not broiler chicks [43, 47]. Multiple studies investigating orexin in avian species theorize that the peptide appears to be more involved in the regulation of energy metabolism than feed intake and sleep/wake cycles [39, 40, 48].

4. Orexin system in avian muscle

Studies investigating the orexin-producing neurons in the rat/mouse brain report that central administration of the neuropeptide facilitates changes in

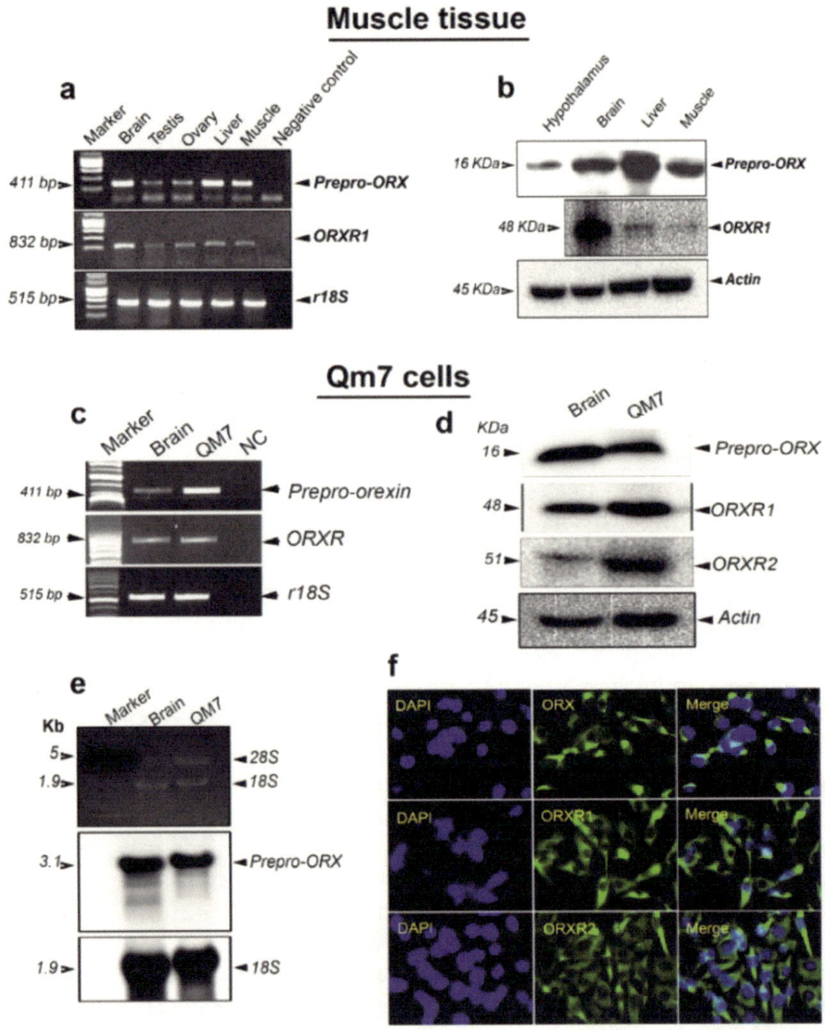

Figure 1.
Prepro-orexin and its related receptors are expressed in broiler chicken muscle (a, b) and QM7 cell line (c–f). (a) RT-PCR. Total RNA (1 μg) was reverse transcribed and subjected to RT-PCR. Brain, testis, and ovary were used as positive controls. (b) Western blot. 70 μg total protein extracted from each tissue were electrophoresed and blotted onto polyvinylidene difluoride membrane. Prepro-orexin (ORX) and orexin receptor 1 (ORXR1) expression was detected by immunoblot using rabbit anti-mouse ORX and rabbit anti-rat ORXR1 antibodies. Hypothalamus and brain were used as positive controls. The figure is a representative picture from one animal. (c) RT-PCR. Total RNA (1 μg) was isolated from chicken brain and QM7 cells and subjected to RT-PCR. (d) Western blot. Total protein extracted from the cells were electrophoresed and blotted onto polyvinylidene difluoride membrane. Prepro-orexin (ORX) and ORXR1/2 expression was detected by immunoblot using rabbit anti-mouse ORX and rabbit anti-rat ORXR1/2 antibodies. Brain was used as positive control. (e) Northern blot. Total RNA (10 μg) was separated by agarose gel electrophoresis and transferred to a nylon membrane and hybridized with specific biotin-labeled DNA probe toward chicken ORX, and 18S. Hybridization signals were detected by FluorChem M MultiFluor system. (f) Immunofluorescence staining. Intracellular ORX system distribution visualized by fluorescent microscope in the presence of a secondary antibody conjugated with Alexa Fluor 488 (green) and DAPI (blue).

skeletal muscle tone [49, 50] in addition to increasing glucose metabolism and insulin sensitivity of skeletal muscle [51]. This indicates that orexin signaling in the central nervous system regulates muscle glucose metabolism by activating muscle sympathetic nerves. Even though the orexin system has been identified in the peripheral tissues of a variety of vertebrate species [52], very few studies have been conducted to determine whether orexin and its related receptors are expressed in vertebrate skeletal muscle and if so, what effects it's presence may have on vertebrate muscle function and physiology. Review of the current literature shows that to date orexin has only been identified as being expressed in the heart muscle of zebrafish [53] and skeletal muscle of goldfish, chicken, and quail [54–56].

We have conducted studies using RT-PCR and Western blot analysis to identify both orexin and its related receptors as being expressed in broiler chicken muscle (**Figure 1a, b**). In addition, RT-PCR, Western blot analysis, Northern blot analysis, and immunofluorescence have been used to illustrate that orexin and its receptors are also expressed in the cytoplasmic compartment of a spontaneously immortalized quail muscle (QM7) cell line (**Figure 1c–f**). In humans, scientific evidence indicates that orexin is a secretory peptide due to its presence in the blood circulation. Additional support for orexin being secreted is given by the first 33 amino acids of human prepro-orexin containing a hydrophobic core followed by residues with small polar side chains, which are characteristics of a secretory signal sequence [57]. Cell culture experiments using the QM7 cell line have been carried out in order to determine whether orexin is also secreted in avian muscle. When QM7 cells are incubated in serum-free growth media in the presence of recombinant human orexin B (rORX-B), an increase in the level of prepro-orexin in the growth media is evident as seen in **Figure 2a**. Further support for the secretion of prepro-orexin is illustrated in **Figure 2b** where the levels of prepro-orexin in the serum-free growth media accumulate over a 72 h period, and in **Figure 2c** where application of the anti-secretory compound brefeldin A causes a buildup of prepro-orexin in the QM7 cell lysate and the subsequent absence of the peptide in the growth media. The expression and secretion of prepro-orexin from avian muscle cells suggests that avian orexin could be a myokine that probably functions in autocrine, paracrine, and/or endocrine roles.

Subsequent experiments were conducted in order to determine whether the orexin system is capable of self-regulation. The effects of 10 and 100 nM of recombinant human orexin A (rORX-A) and rORX-B on mRNA and protein expression of ORX and its related receptors ORXR1 and ORXR2 in QM7 cells are shown in **Figure 2d–g**. A 24 h treatment with either 10 or 100 nM rORX-A upregulated ORX and ORXR1, but not ORXR2 mRNA expression (**Figure 2f**). Treating cells with rORX-B downregulated ORX and ORXR2 and increased ORXR1 mRNA levels (**Figure 2g**). Protein expression levels of ORX, ORXR1, and ORXR2 showed the same patterns as their corresponding genes (**Figure 2d, e**) with the effects on expression levels appearing to be dose-dependent. The ability of rORX-A and rORX-B to differentially regulate gene and protein expression of the orexin system supports the idea that orexins function in an autocrine, paracrine, and/or endocrine role in avian muscle. Although the underlying mechanisms behind the differential regulation of the orexin system is still unknown, the divergent effects of rORX-A and rORX-B on orexin system expression might be related to their structure (presence of disulfide bonds in orexin A and not in orexin B) and their different binding affinity to ORXR2, since they had similar effects on ORXR1 expression.

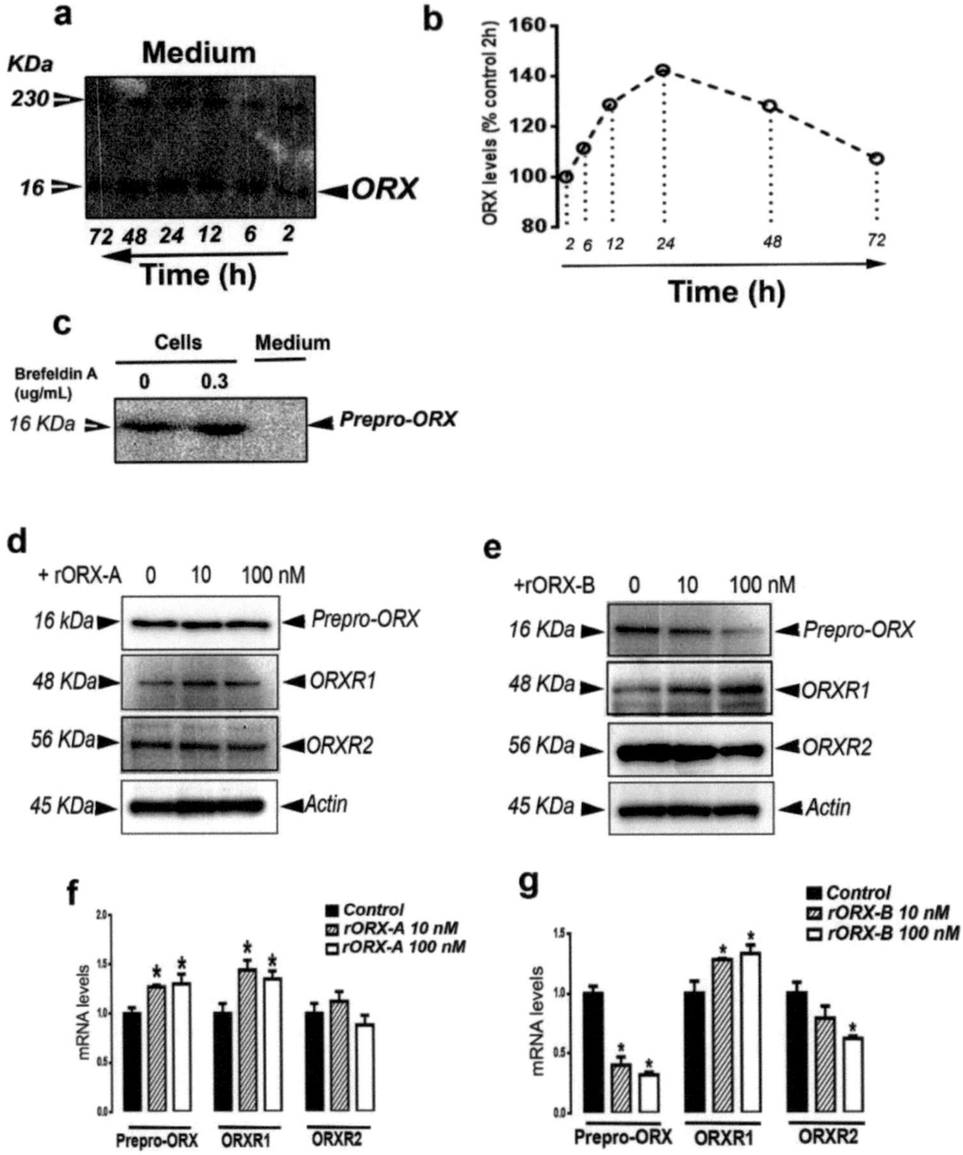

Figure 2.
*Secretion of orexin by QM7 cells (a–c) and effect of orexin treatment on orexin system expression in QM7 cells (d–g). (a) Cell monolayers were incubated in serum-free medium with rORX-B (100 nM) for up to 72 h prior to Western blotting. (b) Percent change in ORX levels when compared to control (2 h post-treatment). (c) Cells were incubated in serum-free medium with or without brefeldin A (0.3 µg/ml) for 12 h. Medium and/or cell lysates were subjected to immunoblot analysis using an orexin antibody. (d–g) Cells were treated with 0 (control), 10, or 100 nM of recombinant orexin A and B for 24 h. Total protein and RNA were isolated and relative expression of ORX and ORXR1/2 was determined. Protein levels were measured by Western blot analysis (d, e). mRNA expression was measured by QPCR using $2^{-\Delta\Delta Ct}$ method (f, g). Data are expressed as means ± SE (n = 6). *Significant difference between orexin-treated and control cells (P < 0.05).*

4.1 The orexin system differentially regulates mitochondrial-related genes and mitochondrial bioenergetics in avian muscle cells

In mammals, orexin has been shown to induce differentiation of brown adipose tissue (BAT), subsequently leading to thermogenesis [58, 59]. One of the effects orexin has in this process is the regulation of genes involved in mitochondrial biogenesis. The treatment of mouse preadipocytes with ORX-A revealed a number of changes in mitochondrial dynamics, including up-regulation of the expression of a number of genes involved in mitochondrial biogenesis (i.e., PGC-1α, PGC-1β, PPARγ1, and UCP1) [58]. These findings were further supported when subsequent immunofluorescence staining revealed an increase in mitochondrial abundance of the treated cells. Studies using other cell types treated with ORX-A have also shown effects on mitochondrial function. Human neuroblastoma cells treated with ORX-A had increased mitochondrial membrane potential [60]. Additionally, in studies using human hepatoma cells [61] and human embryonic kidney cells [62], treatment with ORX-A resulted in increased ATP production that shifted from glycolysis in the cytoplasm to oxidative phosphorylation in the mitochondria. Taken together, these studies indicate that orexin is able to enhance mitochondrial function, biogenesis, and ATP production in mammalian vertebrates.

Similar experiments were carried out by treating QM7 cells with recombinant human orexin, as previously described, in order to observe the differential effects on mitochondrial-related genes, transcriptional regulators, and bioenergetics. **Figure 3** illustrates how orexin causes differential expression of mitochondrial genes. rORX-A had no effect on av-UCP mRNA abundance (**Figure 3b**), but it downregulated the expression of av-ANT (mRNA and protein levels), Ski, and NRF-1 in a dose-dependent manner (**Figure 3a, c, d**). rORX-B, however, down-regulated the expression of av-UCP and increased the expression of Ski and NRF-1 without altering the expression of av-ANT (**Figure 3a–d**). The mitochondrial transcriptional regulators related to these genes were also differentially regulated following treatment with recombinant orexins as seen in **Figure 3e–h**. Both doses of rORX-A caused a significant downregulation of PPARδ and FoxO-1 expression (**Figure 3g, h**). A high dose of rORX-A significantly downregulated the expression of PGC-1β, and both doses did not alter PGC-1α mRNA abundance (**Figure 3e, f**). rORX-B, however, induced the expression of these transcription factors in a dose-dependent manner, but the effects were statistically significant only for PGC-1α, PGC-1β, and FoxO-1 with the high dose (**Figure 3e, f, h**).

Since mitochondria are the powerhouse of the cell with central importance for producing more than 90% of the ATP needed to carry out essential cellular functions, these organelles are important for proper growth and development of skeletal muscle tissue. The changes in expression of mitochondrial-related genes and their transcriptional regulators suggest that orexin may control mitochondrial respiratory function in avian muscle. To gain better insight into the physiological roles of orexins in avian muscle, mitochondrial bioenergetics in QM7 cells treated with 10 and 100 nM of rORX-A and rORX-B were assessed by monitoring basal oxygen consumption rate (OCR) followed by sequential treatment of cells with oligomycin, FCCP (carbonyl cyanide-4-phenylhydrazone), and antimycin A as shown in **Figure 4f**. As described previously [63], the decrease in OCR following oligomycin (which blocks ATP synthase) reveals OCR attributed to ATP synthesis activity. Maximal OCR is revealed in response to the uncoupling compound FCCP, and the difference between maximal OCR and basal OCR (prior to oligomycin) represents mitochondrial oxygen reserve capacity that cells can draw upon when increased energy production is needed. Oxygen consumption that remains following treatment with the electron transport inhibitor antimycin A is attributed to non-mitochondrial OCR (i.e., OCR due to activities other than non-mitochondrial c oxidase activity, such as mitochondrial reactive oxygen species production, oxidase activities, etc.). The amount of OCR attributed to proton leak is determined by the difference between oligomycin and antimycin A-inhibited OCR. When the non-mitochondrial component of cellular OCR was subtracted and by setting maximal OCR following FCCP at 100%, the effects of ORX-A and ORX-B on ATP synthesis, reserve capacity, and proton leak were determined. ATP synthesis was slightly elevated by both orexins, but the effect was not statistically discernable. Analysis of reserve capacity indicated no effect of both doses of rORX-A

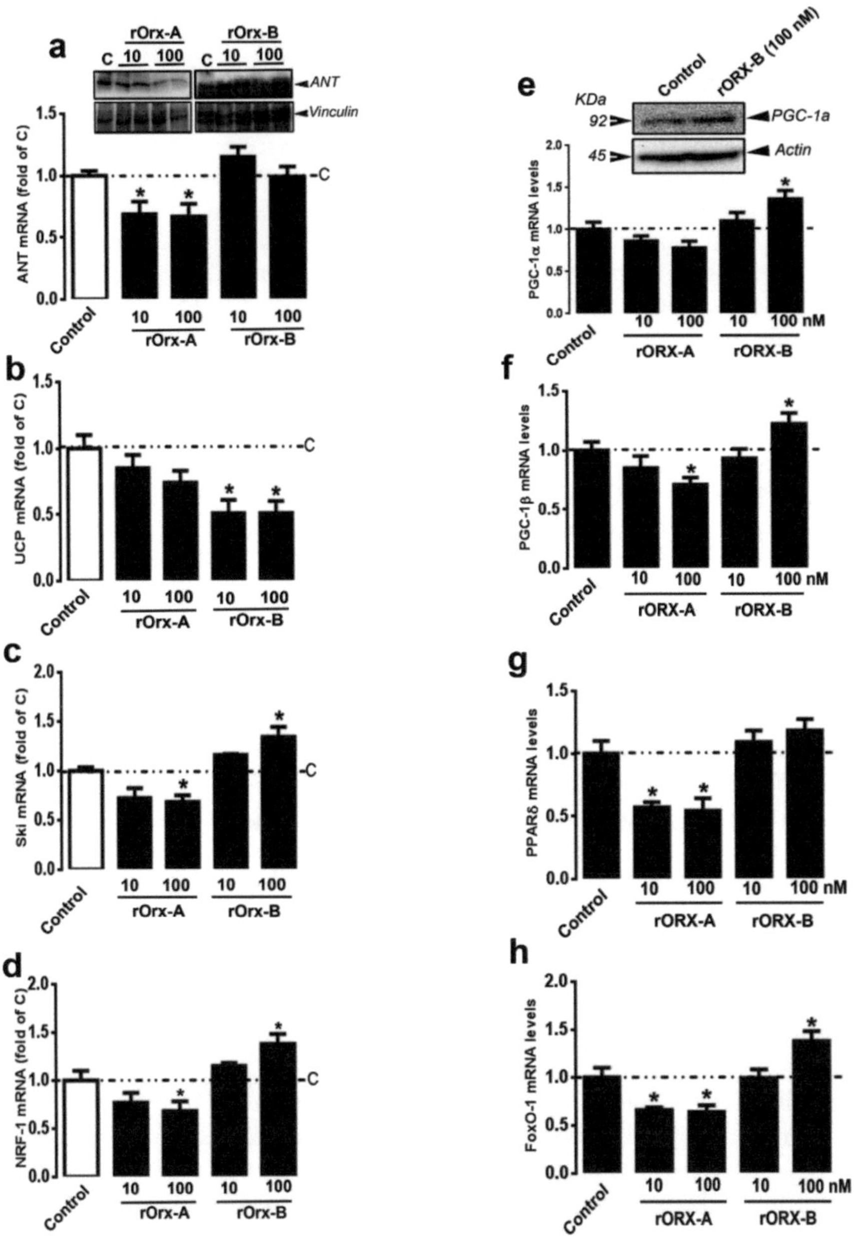

Figure 3.
*Effect of orexin treatment on mitochondrial-related genes (a–d) and mitochondrial-transcriptional regulators (e–h) in QM7 cells. Cells were treated with recombinant orexin A or B (10 and 100 nM) for 24 h and the relative abundance of avian (av)-adenosine nucleotide translocator (ANT; a), UCP (b), Ski (c), nuclear respiratory factor 1 (NRF-1; d), PGC-1α (e), PGC-1β (f), peroxisome proliferator-activated receptor δ (PPARδ; g), and FoxO-1 (h) were determined by QPCR. Untreated cells were used as control. Protein levels of av-ANT and PGC-1α were measured by Western blot analysis. Data are expressed as means ± SE (n = 6). *Significant difference between orexin-treated and control cells (P < 0.05).*

and rORX-B; however, proton leak was significantly decreased by 10 nM of rORX-A, and by 100 nM of rORX-B (**Figure 4g**). The combined data from these experiments illustrate that orexins lead to the alteration of mitochondrial-related genes, their transcriptional regulators, and respiratory function in avian muscle cells. These changes suggest that orexin might also control mitochondrial dynamics (i.e., fusion/fission of mitochondria) in avian muscle.

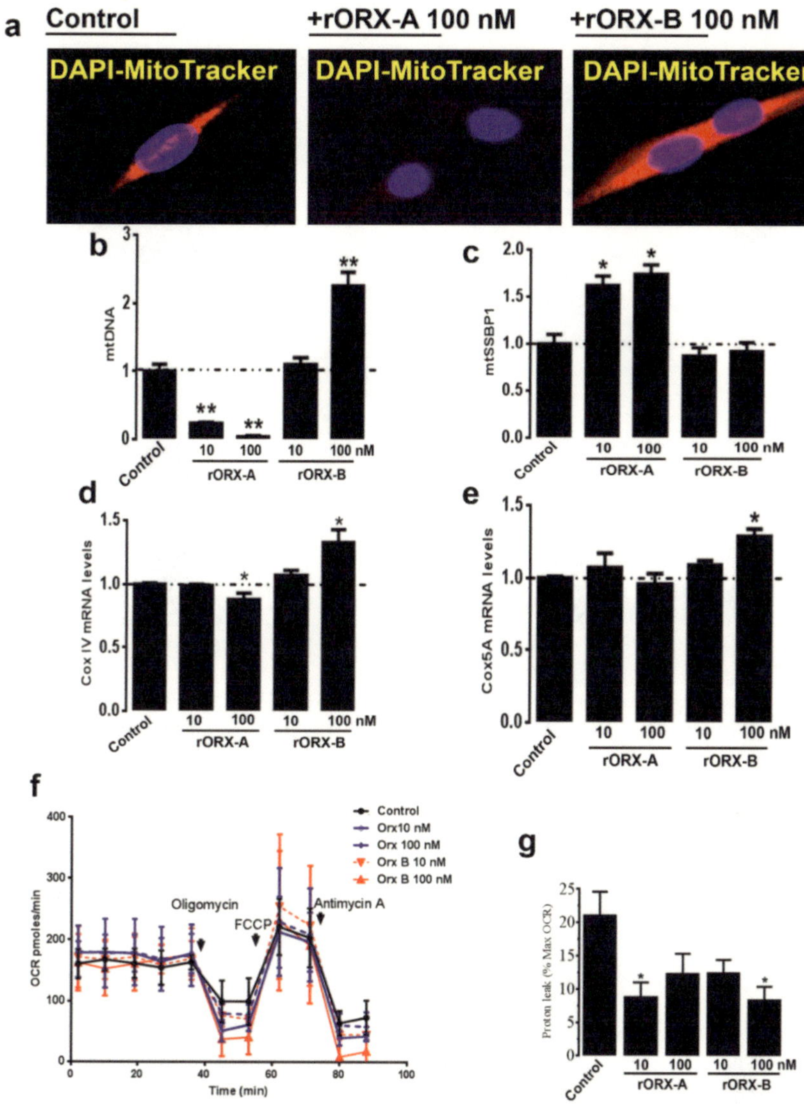

Figure 4.
*Effect of orexin treatment on mitochondrial distribution (a), mitochondrial DNA and mass (b–e), and mitochondrial bioenergetics (f, g) in QM7 cells. a: Cells were cultured in chamber slides and treated with 100 nM of rORX-A or rORX-B for 24 h. Mitochondria were visualized with MitoTracker Red CMX Ros (75 nM) under a fluorescent microscope. Representative images acquired and deconvoluted are shown. (b–e) QM7 cells were treated with orexins (10 and 100 nM) for 24 h. The levels of mtDNA (b) and the relative expression of mtSSBP1 (c), mitochondrial markers CoxIV (d), and Cox5a (e) were determined by real-time PCR. (f, g) Oxygen consumption rate (OCR) (f) and the percentage of OCR due to proton leak (g) was determined using an XF24 flux Analyzer (Agilent Technologies). The values represent the means ±SE (n = 6). *Significant difference between orexin-treated and control cells (P < 0.05).*

4.2 Orexins differentially regulate mitochondrial biogenesis and dynamics in avian muscle cells

Since mtDNA replication and quantitation are a necessary component of mitochondrial biogenesis, the expression of mtDNA and mtSSBP1 was measured in orexin-treated QM7 cells as shown in **Figure 4b, c**. In contrast to rORX-A, in which both doses significantly downregulated mtDNA and upregulated mtSSBP1 expression, rORX-B (high dose) significantly increased mtDNA expression without

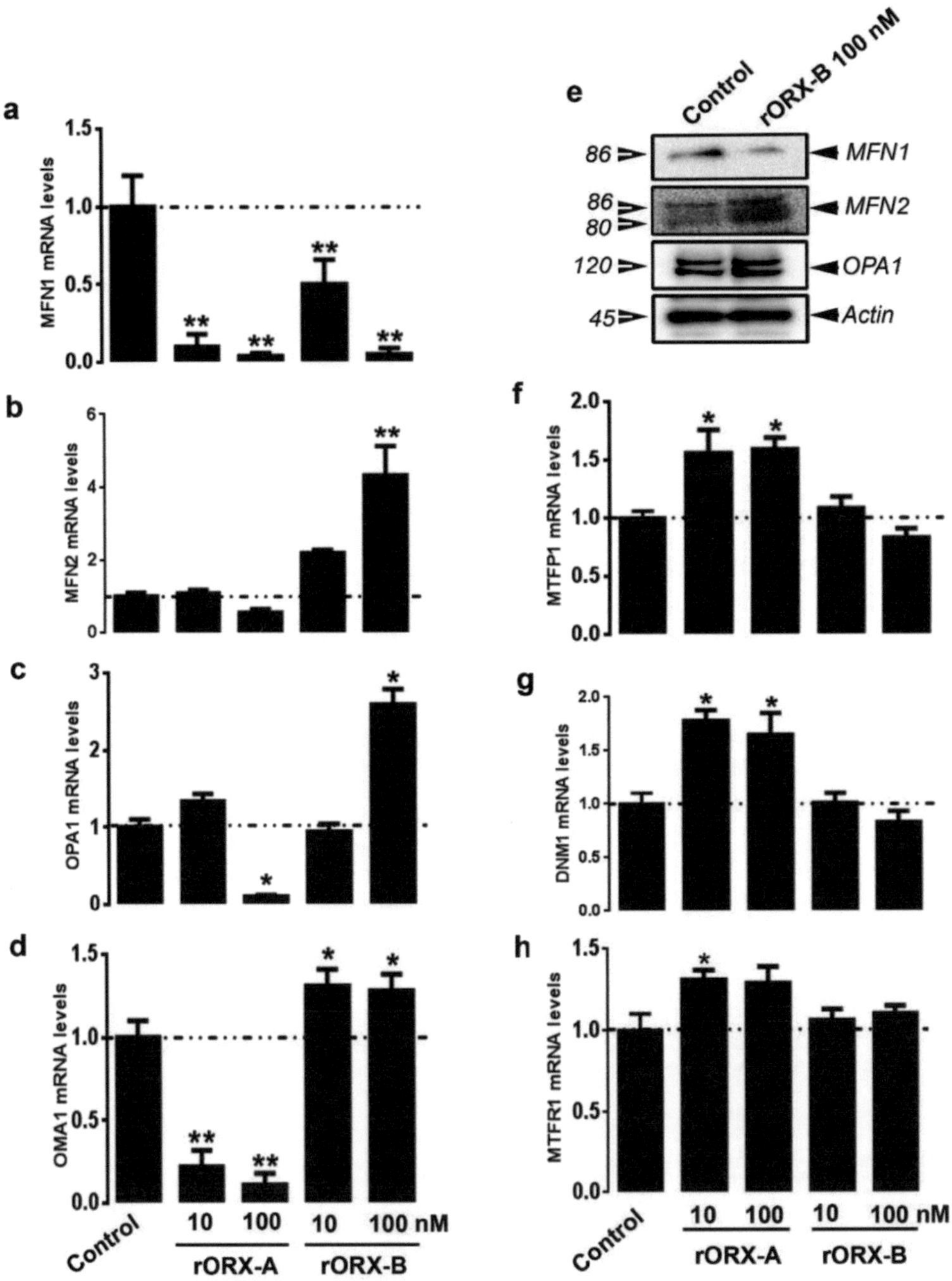

Figure 5.
*Effect of orexin treatment on mitochondrial dynamics-related genes in QM7 cells. QM7 cells were treated with orexins (10 and 100 nM) for 24 h. The relative expression of four genes involved in mitochondrial fusion, MFN1 (a), MFN2 (b), OPA1 (c), OMA1 (d) and three genes involved in mitochondrial fission, MTFP1 (f), DNM1 (g), and MTFR1 (h) was determined by real-time PCR. The protein levels of MFN1, MFN2, and OPA1 were determined by Western blot analysis (e). The values represent the means ±SE (n = 6). *Significant difference between orexin-treated and control cells (P < 0.05).*

affecting mtSSBP1 levels. Consistent with these observations and in contrast to rORX-A, rORX-B increased mitochondrial content as visualized with MitoTracker Red probe staining (**Figure 4a**). Neither rORX-A nor rORX-B affected the expression of the mitochondrial transcription factor TFAM (data not shown). The expression of Cox IV and Cox 5a genes, commonly used markers for mitochondrial mass and biogenesis, was determined. The high dose (100 nM) of rORX-A decreased Cox IV gene expression; however, the high-dose of rORX-B significantly increased Cox IV and Cox 5a mRNA levels compared with untreated cells (**Figure 4d, e**).

Mitochondria are dynamic organelles in the cell that constantly fuse and divide, forming constantly changing tubular networks, according to the needs of the organism, thus leading to alterations in their morphology and function [64]. Since orexin is shown to alter expression of mitochondrial-related genes and transcriptional regulators, it may also control avian muscle mitochondrial dynamics.
The molecular mechanisms that control mitochondrial dynamics are complex and require participation and coordination of both the nuclear and mitochondrial genomes. In mammals this network has been partially unraveled after the identification of some of the genes responsible for mitochondrial fusion [mitofusins (MFN1 and MFN2), and optic atrophy 1 (OPA1)] and fission [dyanamin-related protein 1 (Drp1 or DNM1), fission 1 (FIS1), and mitochondrial protein 18 kDa]. Up until the current study, the integration of such a mitochondrial network is unknown in avian species. Following orexin treatment, the expression of four genes related to mitochondrial fusion and three genes related to mitochondrial fission were measured as shown in **Figure 5**. Recombinant ORX-B at high dose significantly induced the expression of MFN2, OPA1, and OMA1, but decreased the mRNA levels of MFN1 (**Figure 5a–d**). The same effect was observed at the protein levels (**Figure 5e**). However, rORX-A significantly downregulated the expression of MFN1 and OMA1 with both doses, and OPA1 with the high dose, but did not affect that of MFN2 (**Figure 5a–d**). Interestingly, and in contrast to rORX-B, where no significant effects were observed, rORX-A upregulated the expression of mitofission-related genes MTFP1, DNM1, and MTFR1 (**Figure 5f–h**). These orexin-induced changes in the expression of dynamics-related genes may serve in regulating mitochondrial metabolism in muscle cells in response to the needs of the animal during stages of growth and development.

5. Conclusions

Orexins are originally identified as hypothalamic neuropeptides that have potent orexigenic effects on appetite and feeding behavior in mammals. In avian species, however, orexins seem to be myokines that regulate the expression of their own system as well as muscle mitochondrial function, biogenesis, bioenergetics, and dynamics. As intensive genetic selection for fast growth rate, driven by human nutritional needs and economic demands, have resulted in dramatic increase in chicken body weight arising mainly from increased muscle mass, these finding open new vistas on the role of orexin system in muscle development and energy metabolism. Further in depth investigations are warranted to understand the relationship between orexin system, mitochondrial network, and muscle growth and development which, in turn, will be beneficial to the poultry industry as it has the potential to lead to increased production efficiency and reduced economic costs.

Author details

Kentu Lassiter* and Sami Dridi
Department of Poultry Science, Center of Excellence for Poultry Science, University of Arkansas, Fayetteville, Arkansas

*Address all correspondence to: klassit@uark.edu

References

[1] de Lecea L, Kilduff TS, Peyron C, Gao X-B, Foye PE, Danielson PE, et al. The hypocretins: Hypothalamus-specific peptides with neuroexcitatory activity. Proceedings of the National Academy of Sciences of the United States of America. 1998;**95**:322-327. DOI: 10.1073/pnas.95.1.322

[2] Sakurai T, Amemiya A, Ishii M, Matsuzaki I, Chemelli RM, Tanaka H, et al. Orexins and orexin receptors: A family of hypothalamic neuropeptides and G protein-coupled receptors that regulate feeding behavior. Cell. 1998;**92**:573-585. DOI: 10.1016/S0092-8674(00)80949-6

[3] Sakurai T. Orexins and orexin receptors: Implication in feeding behavior. Regulatory Peptides. 1999;**85**:25-30. DOI: 10.1016/S0167-0115(99)00076-2

[4] Chemelli RM, Willie JT, Sinton CM, Elmquist JK, Scammell T, Lee C, et al. Narcolepsy in orexin knockout mice: Molecular genetics of sleep regulation. Cell. 1999;**98**:437-451. DOI: 10.1016/S0092-8674(00)81973-X

[5] Belle MD, Hughes AT, Bechtold DA, Cunningham P, Pierucci M, Burdakov D, et al. Acute suppressive and long-term phase modulation actions of orexin on the mammalian circadian clock. The Journal of Neuroscience. 2014;**34**:3607-3621. DOI: 10.1523/JNEUROSCI.3388-13.2014

[6] Tsuneki H, Murata S, Anzawa Y, Soeda Y, Tokai E, Wada T, et al. Age-related insulin resistance in hypothalamus and peripheral tissues of orexin knockout mice. Diabetologia. 2008;**51**:657-667. DOI: 10.1007/s00125-008-0929-8

[7] Tsuneki H, Wada T, Sasaoka T. Role of orexin in the regulation of glucose homeostasis. Acta Physiologica (Oxford, England). 2010;**198**:335-348. DOI: 10.1111/j.1748-1716.2009.02008.x

[8] Tsuneki H, Wada T, Sasaoka T. Role of orexin in the central regulation of glucose and energy homeostasis. Endocrine Journal. 2012;**59**:365-374. DOI: 10.1507/endocrj.EJ12-0030

[9] Shen Y, Zhao Y, Zheng D, Chang X, Ju S, Guo L. Effects of orexin A on GLUT4 expression and lipid content via MAPK signaling in 3T3-L1 adipocytes. The Journal of Steroid Biochemistry and Molecular Biology. 2013;**138**:376-383. DOI: 10.1016/j.jsbmb.2013.07.005

[10] Skrzypski M, Le TT, Kaczmarek P, Pruszynska-Oszmalek E, Pietrzak P, Szczepankiewicz D, et al. Orexin A stimulates glucose uptake, lipid accumulation and adiponectin secretion from 3T3-L1 adipocytes and isolated primary rat adipocytes. Diabetologia. 2011;**54**:1841-1852. DOI: 10.1007/s00125-011-2152-2

[11] Ciriello J, Li Z, de Oliveira CV. Cardioacceleratory responses to hypocretin-1 injections into rostral ventromedial medulla. Brain Research. 2003;**991**:84-95. DOI: 10.1016/j.brainres.2003.08.008

[12] Zhang W, Fukuda Y, Kuwaki T. Respiratory and cardiovascular actions of orexin-A in mice. Neuroscience Letters. 2005;**385**:131-136. DOI: 10.1016/j.neulet.2005.05.032

[13] Samson WK, Bagley SL, Ferguson AV, White MM. Hypocretin/orexin type 1 receptor in brain: Role in cardiovascular control and the neuroendocrine response to immobilization stress. American Journal of Physiology. Regulatory, Integrative and Comparative Physiology. 2007;**292**:R382-R387. DOI: 10.1152/ajpregu.00496.2006

[14] Watt Executive Guide to World Poultry Trends [Internet]. 2017. Available from: http://www.poultrytrends.com/201711/index.

php?startid=9#/1 [Accessed: 7 December 2018]

[15] Havenstein GB, Ferket PR, Qureshi MA. Growth, livability, and feed conversion of 1957 versus 2001 broilers when fed representative 1957 and 2001 broiler diets. Poultry Science. 2003;**82**:1500-1508. DOI: 10.1093/ps/82.10.1500

[16] Barbut S, Sosnicki AA, Lonergan SM, Knapp T, Ciobanu DC, Gatcliffe LJ, et al. Progress in reducing the pale, soft and exudative (PSE) problem in pork and poultry meat. Meat Science. 2008;**79**:46-63. DOI: 10.1016/j.meatsci.2007.07.031

[17] Guernec A, Berri C, Chevalier B, Wacrenier-Cere N, Le Bihan-Duval E, Duclos MJ. Muscle development, insulin-like growth factor-I and myostatin mRNA levels in chickens selected for increased breast muscle yield. Growth Hormone & IGF Research. 2003;**13**:8-18. DOI: 10.1016/S1096-6374(02)00136-3

[18] Krzysik-Walker SM, Ocon-Grove OM, Maddineni SR, Hendricks GL 3rd, Ramachandran R. Is visfatin an adipokine or myokine? Evidence for greater visfatin expression in skeletal muscle than visceral fat in chickens. Endocrinology 2008;**149**: 1543-1550. DOI: 10.1210/en.2007-1301

[19] Akiba Y, Chida Y, Takahashi T, Ohtomo Y, Sato K, Takahashi K. Persistent hypoglycemia induced by continuous insulin infusion in broiler chickens. British Poultry Science. 1999;**40**:701-705. DOI: 10.1080/00071669987124

[20] Seki Y, Sato K, Kono T, Abe H, Akiba Y. Broiler chickens (Ross strain) lack insulin-responsive glucose transporter GLUT4 and have GLUT8 cDNA. General and Comparative Endocrinology. 2003;**133**:80-87. DOI: 10.1016/S0016-6480(03)00145-X

[21] Barre H, Cohen-Adad F, Duchamp C, Rouanet JL. Multilocular adipocytes from muscovy ducklings differentiated in response to cold acclimation. The Journal of Physiology. 1986;**375**:27-38. DOI: 10.1113/jphysiol.1986.sp016103

[22] Hood RL. The cellular basis for growth of the abdominal fat pad in broiler-type chickens. Poultry Science. 1982;**61**:117-121. DOI: 10.3382/ps.0610117

[23] Furuse M, Ando R, Bungo T, Ao R, Shimojo M, Masuda Y. Intracerebroventricular injection of orexins does not stimulate food intake in neonatal chicks. British Poultry Science. 1999;**40**:698-700. DOI: 10.1080/00071669987115

[24] Stadel JM, Wilson S, Bergsma D. Orphan g protein-coupled receptors: A neglected opportunity for pioneer drug discovery. Trends in Pharmacological Sciences. 1997;**18**:430-437. DOI: 10.1016/S0165-6147(97)01117-6

[25] Shibahara M, Sakurai T, Nambu T, Takenouchi T, Iwaasa H, Egashira SI, et al. Structure, tissue distribution, and pharmacological characterization of xenopus orexins. Peptides. 1999;**20**:1169-1176. DOI: 10.1016/S0196-9781(99)00120-5

[26] Alvarez CE, Sutcliffe JG. Hypocretin is an early member of the incretin gene family. Neuroscience Letters. 2002;**324**:169-172. DOI: 10.1016/S0304-3940(02)00195-7

[27] Sakurai T. Reverse pharmacology of orexin: From an orphan GPCR to integrative physiology. Regulatory Peptides. 2005;**126**:3-10. DOI: 10.1016/j.regpep.2004.08.006

[28] Tsujino N, Sakurai T. Orexin/hypocretin: A neuropeptide at the interface of sleep, energy homeostasis, and reward system. Pharmacological Reviews. 2009;**61**:162-176. DOI: 10.1124/pr.109.001321

[29] Soppet DR, Li Y, Rosen CA. Human Genome Sciences Inc. Human Neuropeptide Receptor. 1996. World Patent No. WO9634877

[30] Karteris E, Randeva HS, Grammatopoulos DK, Jaffe RB, Hillhouse EW. Expression and coupling characteristics of the CRH and orexin type 2 receptors in human fetal adrenals. The Journal of Clinical Endocrinology and Metabolism. 2001;**86**:4512-4519. DOI: 10.1210/jcem.86.9.7849

[31] Randeva HS, Karteris E, Grammatopoulos DK, Hillhouse EW. Expression of orexin-A and functional orexin type 2 receptors in the human adult adrenals: Implications for adrenal function and energy homeostasis. The Journal of Clinical Endocrinology and Metabolism. 2001;**86**:4808-4813. DOI: 10.1210/jcem.86.10.7921

[32] Karteris E, Machado RJ, Chen J, Zervou S, Hillhouse EW, Randeva HS. Food deprivation differentially modulates orexin receptor expression and signaling in the rat hypothalamus and adrenal cortex. American Journal of Physiology. Endocrinology and Metabolism. 2005;**288**:E1089-E1100. DOI: 10.1152/ajpendo.00351.2004

[33] Hepler JR, Gilman AG. G proteins. Trends in Biochemical Sciences. 1992;**17**:383-387. DOI: 10.1016/0968-0004(92)90005-T

[34] Gautam N, Downes GB, Yan K, Kisselev O. The g-protein betagamma complex. Cellular Signalling. 1998;**10**:447-455. DOI: 10.1016/S0898-6568(98)00006-0

[35] Kukkonen JP. Physiology of the orexinergic/hypocretinergic system: A revisit in 2012. American Journal of Physiology. Cell Physiology. 2013;**304**(1):C2-C32. DOI: 10.1152/ajpcell.00227.2012

[36] Kukkonen JP, Leonard CS. Orexin/hypocretin receptor signaling cascades. British Journal of Pharmacology. 2014;**171**(2):314-331. DOI: 10.1111/bph.12324

[37] Ohkubo T, Boswell T, Lumineau S. Molecular cloning of chicken prepro-orexin cDNA and preferential expression in the chicken hypothalamus. Biochimica et Biophysica Acta. 2002;**1577**:476-480. DOI: 10.1016/S0167-4781(02)00483-9

[38] Ohkubo T, Tsukada A, Shamoto K. cDNA cloning of chicken orexin receptor and tissue distribution: Sexually dimorphic expression in chicken gonads. Journal of Molecular Endocrinology. 2003;**31**:499-508. DOI: 10.1677/jme.0.0310499

[39] Miranda B, Esposito V, de Girolamo P, Sharp PJ, Wilson PW, Dunn IC. Orexin in the chicken hypothalamus: Immunocytochemical localization and comparison of mRNA concentrations during the day and night, and after chronic food restriction. Brain Research. 2013;**1513**:34-40. DOI: 10.1016/j.brainres.2013.03.036

[40] Godden KE, Landry JP, Slepneva N, Migues PV, Pompeiano M. Early expression of hypocretin/orexin in the chick embryo brain. PLoS One. 2014;**9**(9):e106977. DOI: 10.1371/journal.pone.0106977

[41] Arcamone N, D'Angelo L, de Girolamo P, Lucini C, Pelagalli A, Castaldo L. Orexin and orexin receptor like peptides in the gastroenteric tract of gallus domesticus: An immunohistochemical survey on presence and distribution. Research in Veterinary Science. 2014;**96**:234-240. DOI: 10.1016/j.rvsc.2014.02.002

[42] Edwards CM, Abusnana S, Sunter D, Murphy KG, Ghatei MA, Bloom SR. The effect of the orexins on food intake: Comparison with

neuropeptide Y, melanin-concentrating hormone and galanin. The Journal of Endocrinology. 1999;**160**:R7-R12. DOI: 0022-0795/99/0160-00R7

[43] Katayama S, Hamasu K, Shigemi K, Cline MA, Furuse M. Intracerebroventricular injection of orexin-A, but not orexin-B, induces arousal of layer-type neonatal chicks. Comparative Biochemistry and Physiology A. 2010;**157**:132-135. DOI: 10.1016/j.cbpa.2010.05.018

[44] da Silva ES, dos Santos TV, Hoeller AA, dos Santos TS, Pereira GV, Meneghelli C, et al. Behavioral and metabolic effects of central injections or orexins/hypocretins in pigeons (*Columba livia*). Regulatory Peptides. 2008;**147**:9-18. DOI: 10.1016/j.regpep.2007.12.003

[45] Phillips-Singh D, Li Q, Takeuchi S, Ohkubo T, Sharp PJ, Boswell T. Fasting differentially regulates expression of agouti-related peptide, pro-opiomelanocortin, prepro-orexin, and vasoactive intestinal polypeptide mRNAs in the hypothalamus of Japanese quail. Cell and Tissue Research. 2003;**313**: 217-225. DOI: 10.1007/s00441-003-0755-8

[46] Song Z, Liu L, Yue Y, Jiao H, Lin H, Sheikhahmadi A, et al. Fasting alters protein expression of AMP-activated protein kinase in the hypothalamus of broiler chicks (*Gallus gallus domesticus*). General and Comparative Endocrinology. 2012;**178**:546-555. DOI: 10.1016/j.ygcen.2012.06.026

[47] Katayama S, Shigemi K, Cline MA, Furuse M. Clorgyline inhibits orexin-A-induced arousal in layer-type chicks. The Journal of Veterinary Medical Science. 2011;**73**:471-474. DOI: 10.1292/jvms.10-0358

[48] Song Z, Everaert N, Wang Y, Decuypere E, Buyse J. The endocrine control of energy homeostasis in chickens. General and Comparative Endocrinology. 2013;**190**:112-117. DOI: 10.1016/j.ygcen.2013.05.006

[49] Kiyashchenko LI, Mileykovskiy BY, Lai Y-Y, Siegel JM. Increased and decreased muscle tone with orexin (hypocretin) microinjections in the locus coeruleus and pontine inhibitory area. Journal of Neurophysiology. 2001;**85**:2008-2016. DOI: 10.1152/jn.2001.85.5.2008

[50] Mileykovskiy BY, Kiyashchenko LI, Siegel JM. Muscle tone facilitation and inhibition after orexin-A (hypocretin-1) microinjections into the medial medulla. Journal of Neurophysiology. 2002;**87**:2480-2489. DOI: 10.1152/jn.2002.87.5.2480

[51] Shiuchi T, Haque MS, Okamoto S, Inoue T, Kageyama H, Lee S, et al. Hypothalamic orexin stimulates feeding-associated glucose utilization in skeletal muscle via sympathetic nervous system. Cell Metabolism. 2009;**10**: 466-480. DOI: 10.1016/j.cmet.2009.09.013

[52] Wong KK, Ng SY, Lee LT, Ng HK, Chow BK. Orexins and their receptors from fish to mammals: A comparative approach. General and Comparative Endocrinology. 2011;**171**:124-130. DOI: 10.1016/j.ygcen.2011.01.001

[53] Kaslin J, Nystedt JM, Ostergard M, Peitsaro N, Panula P. The orexin/hypocretin system in zebrafish is connected to the aminergic and cholinergic systems. The Journal of Neuroscience. 2004;**24**:2678-2689. DOI: 10.1523/JNEUROSCI.4908-03.2004

[54] Matsuda K, Azuma M, Kang KS. Orexin system in teleost fish. In: Litwack G, editor. Vitamins and Hormones. 1st ed. San Diego: Elsevier; 2012. pp. 341-361. DOI: 10.1016/B978-0-12-394623-2.00018-4

[55] Lassiter K, Greene E, Piekarski A, Faulkner OB, Hargis BM, Bottje W,

et al. Orexin system is expressed in avian muscle cells and regulates mitochondrial dynamics. American Journal of Physiology. Regulatory, Integrative and Comparative Physiology. 2015;**308**:R173-R187. DOI: 10.1152/ajpregu.00394.2014

[56] Nguyen PH, Greene E, Kong BW, Bottje W, Anthony N, Dridi S. Acute heat stress alters the expression of orexin system in quail muscle. Frontiers in Physiology. 2017;**8**:1079. DOI: 10.3389/fphys.2017.01079

[57] Arihara Z, Takahashi K, Murakami O, Totsune K, Sone M, Satoh F, et al. Immunoreactive orexin-A in human plasma. Peptides. 2001;**22**:139-142. DOI: 10.1016/S0196-9781(00)00369-7

[58] Sellayah D, Bharaj P, Sikder D. Orexin is required for brown adipose tissue development, differentiation, and function. Cell Metabolism. 2011;**14**: 478-490. DOI: 10.1016/j.cmet.2011.08.010

[59] Swami M. Metabolism: Orexin acts on brown fat. Nature Medicine. 2011;**17**:1356. DOI: 10.1038/nm.2563

[60] Pasban-Aliabadi H, Esmaeili-Mahani S, Abbasnejad M. Orexin-A protects human neuroblastoma SH-SY5Y cells against 6-hydroxydopamine-induced neurotoxicity: Involvement of PKC and PI3K signaling pathways. Rejuvenation Research. 2017;**20**:125-133. DOI: 10.1089/rej.2016.1836

[61] Wan X, Liu Y, Zhao Y, Sun X, Fan D, Guo L. Orexin A affects HepG2 human hepatocellular carcinoma cells glucose metabolism via HIF-1α-dependent and -independent mechanism. PLoS One. 2017;**12**(9):e0184213. DOI: 10.1371/journal.pone.0184213

[62] Sikder D, Kodadek T. The neurohormone orexin stimulates hypoxia-inducible factor-1 activity. Genes & Development. 2007;**21**: 2995-3005. DOI: 10.1101/gad.1584307

[63] Hill BG, Awe SO, Vladykovskaya E, Ahmed Y, Liu SQ, Bhatnagar A, et al. Myocardial ischaemia inhibits mitochondrial metabolism of 4-hydroxy-trans-2-nonenal. The Biochemical Journal. 2009;**417**:513-524. DOI: 10.1042/BJ20081615

[64] Chen H, Chan DC. Emerging functions of mammalian mitochondrial fusion and, fission. Human Molecular Genetics. 2005;**14**(2):R283-R289. DOI: 10.1093/hmg/ddi270

Chapter 4

Noncoding RNAs in the Cardiovascular System: Exercise Training Effects

Noemy Pereira, Camila Gatto, Edilamar Menezes de Oliveira and Tiago Fernandes

Abstract

Exercise training (ET) represents a non-pharmacological treatment that can attenuate or even reverse the process of cardiovascular diseases (CVD), by stimulating protein synthesis, angiogenesis, mitochondrial biogenesis, anti-inflammatory, and anti-oxidative effects that are involved to enhance the performance and improved quality of life. Despite the benefits of exercise, the intricacies of their underlying molecular mechanisms remain largely unknown. Noncoding RNAs (ncRNAs) have been recognized as a major regulatory network governing gene expression in several physiological processes and appeared as pivotal modulators in a myriad of cardiovascular processes under physiological and pathological conditions. However, little is known about ncRNA expression and role in response to exercise. Here we review the current understanding of the ncRNA role in exercise-induced adaptations focused on the cardiovascular system and address their potential role in clinical applications for cardiovascular diseases.

Keywords: microRNA, lncRNAs, heart, vessels, exercise and cardiovascular diseases

1. Introduction

Exercise training (ET) has been shown to undergo beneficial changes in both cardiac structure and function, protecting the heart and vessels against the development of cardiovascular diseases (CVD) [1–4]. CVD are a growing cause of morbidity and mortality throughout the world and often develop after a period of abnormal heart growth termed pathological cardiac hypertrophy (CH). The occurrence of risk factors, such as diabetes, hypertension, obesity, and advanced age leads to substantial complications of CVD. Loss of cardiomyocytes and capillaries accompanied by fibrosis contributes to decreased cardiac function ultimately leading to heart failure (HF). Although current management has improved survival in patients with CVD, such therapies do not fully address the underlying cause and, as a result, HF progresses. Thus, understanding the pathways that rescue cardiovascular remodeling (CR) and function could have important clinical implications [3–6].

Genome-wide analyses and ribonucleic acids (RNA) sequencing revealed that a large part of the human genome (~99%) do not encode proteins but are transcriptionally active and give rise to a broad spectrum of noncoding RNAs (ncRNAs).

The ENCODE Project in 2000 (see https://www.encodeproject.org/), that aimed to characterize all the functional elements in the human genome, helped researchers to understand the eukaryotic genome complexity. Although ncRNAs have been considered as junk in the human genome, currently it represent as one of the newest study targets of CVD by potential role in controlling gene expression [7, 8].

Bartolomei et al. [9] discovered the first ncRNA (H19) in mice and it was one of the first maternally imprinted long ncRNAs (lncRNAs) to be identified [9]. Since then, thousands of ncRNAs have been described and are classified into two large groups: small ncRNAs, which are up to 200 nucleotides (nt), and lncRNAs, which are longer than 200 nucleotides. Small ncRNAs comprise microRNAs (miRNAs), PIWI interacting RNAs (piRNAs), transfer RNAs (tRNAs), and small nucleolar RNAs (snoRNAs). Long ncRNAs mainly refers to natural antisense transcripts and other lncRNAs. The discovery of the new class of RNAs increased our knowledge about epigenetic, transcriptional, and translational regulation of gene expression under physiological and pathological conditions. Thus, the possibility of ncRNAs as a potential tool for the treatment of CVD is of great interest [6, 8, 10–12].

ncRNAs such as miRNAs and lncRNAs are regulators of cell development, differentiation, and growth, exert their functions in both nuclear and cytoplasmic compartments. In the cardiovascular system, ncRNAs regulate cardiac development, inflammation, hypertrophy, fibrosis, and regeneration [6, 10–12]. Beyond this, altered miRNA expression can be found in the blood of patients with various CVD [10, 13]. ET is well known to promote cardiovascular profit in which it can vary according to type, intensity, and duration of exercise. However, little is known about the regulatory interaction networks among the multiple classes of RNAs or the mechanisms regulated by exercise-induced ncRNAs on heart and vessel [1, 3, 5, 14]. The contribution of ncRNAs to these adaptations has the potential not only to reveal novel aspects of cardiovascular system but also to identify new targets for prevention and treatment of the CVD.

In this review, we provide an overview of the current knowledge of the cardiovascular effects of ET on ncRNAs, with a specific focus on miRNAs, their role in regulating cardiovascular remodeling (CR) and their potential application for the treatment of CVD. These findings will provide insights that can aid in the development of new therapeutic interventions for these pathologies.

2. miRNAs, a family of small ncRNAs

miRNAs, the most popular class of small ncRNAs, are a class of single-stranded, endogenous, conserved, RNAs (~21–23 nucleotides) that regulate gene expression at the posttranscriptional level. They bind to messenger RNA (mRNA) by base pairing to complementary sequences within the 3′ untranslated region (3′UTR), and cause gene silencing through the inhibition of translation and/or degradation of mRNA [15–17]. However, recent studies have suggested that miRNA-binding sites are also located in 5′ UTRs or open reading frames (ORFs), and the mechanism of miRNA-mediated regulation from these sites has not been identified [18]. miRNA sequences are in diverse regions of the genome. Most canonical miRNAs in humans are encoded within introns of coding or noncoding genes. However, some miRNAs may be encoded by exons or may be associated in the same loci, organized as a polycistronic transcription unit. Small sequences and imperfect complementary, a single miRNA can target several to hundreds of distinct mRNAs. Conversely, an mRNA can be targeted by diverse miRNAs simultaneously, thus miRNAs are estimated to regulate the expression of more than a third of human protein-coding genes [15–17].

The first miRNA, lin-4, was discovered in *C. elegans* in 1993. Lin-4 regulates *C. elegans* development by binding to the lin-14 mRNA to inhibit lin-14 protein expression [19]. Since then, it has been recognized that miRNAs are a highly conserved class of small RNAs present in most organisms and a variety of miRNAs have been discovered suggesting that there are in excess of 2650 mature sequences in human (see http://www.mirbase.org). Studies have demonstrated that miRNAs influence several biological events, including embryogenesis, cell death, differentiation, proliferation, and cell growth. Furthermore, developments in miRNA-related technologies, such as miRNA expression profiling and synthetic oligoRNA, have contributed to the identification of miRNAs involved in a number of physiological and pathological conditions [15, 16, 20].

The biogenesis of the miRNA starts in the nucleus; RNA polymerase II initially transcribes miRNAs into long segments of coding or noncoding RNA, known as pri-miRNAs, which are usually capped and polyadenylated. The pri-miRNAs harbor a local hairpin structure that is then cropped by a nuclear enzyme Drosha and their cofactor Pasha (also known as DGCR8) into pre-miRNAs (~70 nucleotides). After nuclear processing, the pre-miRNA is exported into the cytoplasm by Exportin-5 in a RanGTP-dependent manner. Subsequently, the enzyme Dicer removes the terminal loop of the pre-miRNAs to generate the miRNA:miRNA∗ duplex (dsRNA) of ~20–25 nucleotides. The dsRNA is loaded into the miRNA-associated multiprotein RNA-induced silencing complex (RISC), which includes the Argonaute proteins. One strand of the miRNA is preferentially retained in this complex and becomes

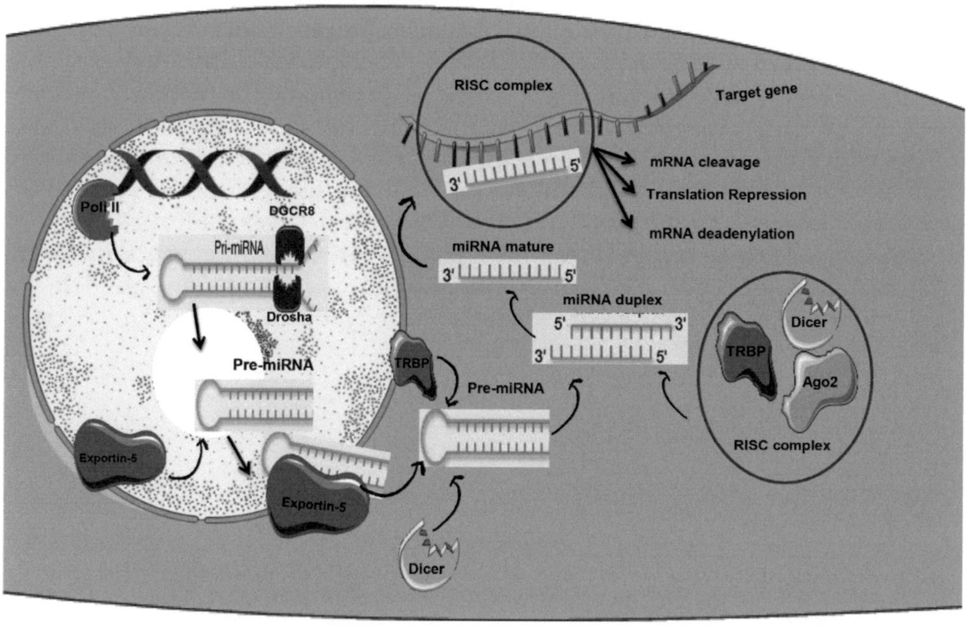

Figure 1.
miRNA biogenesis. miRNA genes are transcribed by RNA polymerase II (Pol II) to generate the primary transcripts (pri-miRNAs). The initial processing of the primary transcript is mediated by the Drosha-DiGeorge syndrome critical region gene 8 (DGCR8) complex (also known as the microprocessor complex) that generates ~70 nucleotide (nt) pre-miRNAs. Pre-miRNA has a short stem plus, which is recognized by the nuclear export factor exportin 5. Once exported from the nucleus, the cytoplasmic RNase III Dicer catalyzes the production of miRNA duplexes. Dicer, TRBP (TAR RNA-binding protein; also known as TARBP2), and Argonaute (AGO) 1–4 mediate the processing of pre-miRNA and the assembly of the RISC (RNA-induced silencing complex). Within this complex, one strand of the miRNA duplex is removed resulting in a single-stranded miRNA, partially complementary to target mRNA, which remains in the complex. miRNA complex interaction with mRNA induces posttranscriptional silencing through as both mRNA destabilization and translational repression (https://smart.servier.com/image-set-download/).

the mature miRNA depending on its thermodynamic stability; the opposite strand, known as the passenger strand, is eliminated from the complex. The RISC acts as a guide by base pairing of miRNA with its target mRNAs, resulting in mRNA cleavage, mRNA deadenylation, and translation repression (**Figure 1**). It is well recognized that pairing in the 'seed' region of the miRNA (nucleotides 2–7 or 8) appears critical for target recognition. Thus, miRNAs that bind to target mRNAs with imperfect complementarity repress target gene expression via translational silencing. In contrast, miRNAs that bind to their target mRNAs with perfect complementarily induce mRNA degradation [15–17, 20].

3. miRNAs regulation by long noncoding RNAs

Emerging evidences suggest that miRNAs may be affected by long noncoding RNAs (lncRNAs) [14, 21]. While there is plenty evidence of miRNAs involvement in exercise adaptations, investigations on lncRNAs are still lacking [14]. LncRNAs are a heterogeneous group of transcripts modulating gene expression at multiple and more complex levels than miRNAs. One way to classify lncRNAs is according to their mechanism of action: signal, decoy, guide, scaffold, enhancer, or sponge lncRNAs (see the review by [22, 23]).

Similar to miRNAs, lncRNAs have been shown to be involved in CVD with several reports about their specific expression in different cardiac diseases [21, 24–30]. In this context, it was recently showed that H19, which is a maternally imprinted lncRNA, is implicated in CVD [24, 31]. H19 is highly expressed during embryogenesis but is strongly repressed after birth, with a significant re-expression of lncRNA H19 in CVD. Liu et al. [24] showed that miRNA-675 mediates the anti-hypertrophic effect of H19 on cardiomyocytes acting as a negative regulator of cardiac hypertrophy. In addition, it was demonstrated that H19 exon1 encodes miR-675-3p and miR-675-5p, which are up-regulated in pathological CH. Prediction algorithms identified a pro-hypertrophic factor, Ca/calmodulin-dependent protein kinase IIδ (CaMKIIδ), as potential targets for miR-675-3p and miR-675-5p, which was confirmed by luciferase reporter gene assays [24]. Moreover, high H19 expression has been linked to hyperhomocysteinemia, a known risk factor for coronary artery disease [32]. Furthermore, H19 was reported to sponge let-7 family miRNAs, which are believed

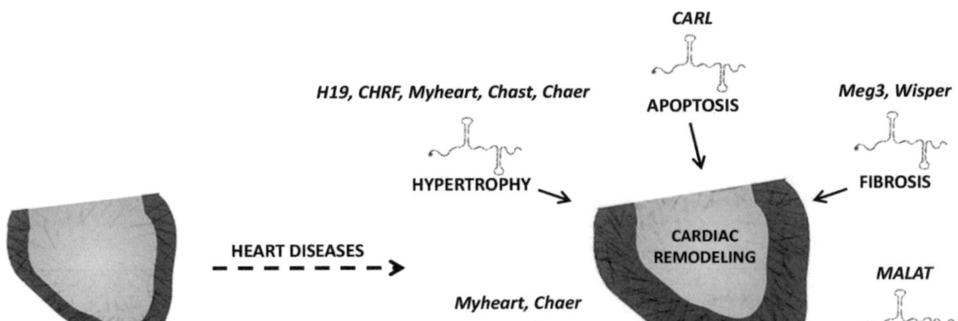

Figure 2.
Cardiac disease-induced cardiac remodeling (CR) is characterized by an aberrant profile of myocardium growth, accompanied by fibrosis, apoptosis, and cardiac dysfunction. lncRNAs H19, CHRF, Myheart, Chast, and Chaer have been described to regulate pathological cardiac hypertrophy; cardiac fibroblast-enriched lncRNAs Meg3 and Wisper controlling cardiac fibrosis; and lncRNAs CHRF, Myheart, and Chaer associated with heart failure in response to cardiac injury.

to have atheroprotective role [21, 33]. Thus, lncRNA-miRNA-mRNA axis reveals a promising therapeutic strategy for CVD.

Cardiac hypertrophy-related factor (CHRF) [26], myosin heavy-chain-associated RNA transcripts (Myheart) [25], cardiac hypertrophy-associated epigenetic regulator (Chaer) [27], cardiac hypertrophy-associated transcript (Chast) [28], cardiac fibroblast-enriched lncRNA maternally expressed 3 (Meg3) [29] and Wisp2 super-enhancer-associated RNA (Wisper) [30] are lncRNAs linked to CR (**Figure 2**). Additionally, recent gain- and loss-of-function studies, both *in vitro* in cell lines and *in vivo* in mouse models, have demonstrated that lncRNAs play critical roles in cardiovascular biology and diseases.

4. Heart, exercise, and miRNAs

4.1 Cardiac remodeling and dysfunction: exercise training effects

The mammalian heart is a muscle and is the first organ to form in the embryo. The development of heart is highly complex and involves the integration of multiple cell types as well as a connection between vascular system and blood vessels. However, its uninterrupted development and function are determinant to organism survival [34, 35]. The primary function of heart is to contract and pump blood throughout the circulatory system to deliver oxygen and nutrients to tissues and to transport carbon dioxide back to the lungs [36, 37]. This organ can be remodeled according to environmental stimuli or external stimuli.

Cardiac remodeling (CR) is an adaptation to increased cardiac overload including a variety of mechanical, hemodynamic, and hormonal factors that promotes changes in structure, dimensions, mass, shape, and functions of the heart as well as cardiac cells [34, 38], and has been generally defined as either physiological (i.e., normal) or pathological (i.e., detrimental) [34]. Physiological growth occurs in response to exercise, pregnancy, or postnatal growth [34] causing heart hypertrophy, which is reversible and characterized by normal cardiac morphology and function [39]. On the other hand, in pathological hypertrophy, contractile performance, for example, can be perturbed or reduced in response to diverse pathophysiological stimuli, which also causes the typically heart remodeling, in association with increases in myocyte cell volume [36, 39]. However, it can slowly progressives to become maladaptive, leading to a decompensation in cardiac function [40].

Cardiac dysfunction is the main consequence of pathological CR, a process that involves a complex series of transcriptional, signaling, structural, electrophysiological, and functional events, occurring within the cardiac myocyte and in the other cellular elements within the ventricle (fibroblasts, endothelial cells, T-cells, etc.). Thus, it starts with genetic changes in response to cardiac insults, with re-expression of fetal genes (atrial and β-type natriuretic peptide, β-myosin heavy chain, and α-skeletal muscle actin), fibrosis, endothelial dysfunction, and inflammation. Consequently, cellular and molecular changes occur, resulting in progressive loss of ventricular function, asymptomatic at first, that evolves to signs and symptoms of HF [41–43].

ET is the most effective non-pharmacological intervention to prevent or reduce cardiovascular disturbances [2, 44]. ET can be classified as static or dynamic and leads to two different types of intermittent chronic cardiac overload, which induces morphological changes in the heart [45]. This changes will depend on factors such as the type of exercise used (aerobic or resistance), volume (training time), intensity (degree of training load), and frequency of ET (number of training session at any given time) [2, 45], and is directly related to maximal aerobic capacity or VO_2 max [42]. It is known that ET is one of major stimuli for physiological cardiac hypertrophy (CH) [46]. The main physiological responses attributed to CR are the increase in contractility, the improvement of the transient Ca^{2+} intracellular, increase in expansion with significant improvements in myocardial oxygenation, and additional endothelium-dependent functions that prevent ischemic events [42].

In contrast to pathological CH that occurs in response to continuous stimulus on heart (arterial hypertension, valve disease, or aortic stenosis, for example), CH induced by ET consists of a frequent, but intermittent stimulus cardiac hemodynamic overload [2]. Thus, a pressure overload in response to resistance training, for example, causes an increase in the width of the cardiac myocytes without dilation of the chamber and causes a concentric left ventricle (LV) hypertrophy [39]. On the other hand, the increase in size of cardiomyocytes by aerobic ET is mainly associated with an eccentric LV hypertrophy phenotype, induced by volume overload, with predominant myocyte growth in series, promoting dilation of the cardiac chamber. Either concentric or eccentric CH by ET is not followed by fibrosis and re-expression of fetal genes, which is linked to preserved or improved cardiac function [39, 44].

Medeiros et al. [47] showed that CH induced by moderate aerobic ET causes resting bradycardia in rats by an increase in cardiac vagal activity, which is associated to increased VO_2 peak [44, 47, 48] and an increase in nitric oxide (NO) and angiogenesis levels in heart of experimental animals. The increased cross-sectional area in the heart contributes to the increase in ventricular systolic volume and cardiac output, which improves aerobic capacity [49]. Thus, ET has been widely used as a preventive therapeutic strategy for prevention or treatment of diseases in cardiovascular system. However, the underlying molecular mechanisms that regulate ET responses have not been fully elucidated, especially with regard to CH [50]. In this sense, emerging evidences have shown that miRNAs may play an important role in both heart development and CR [44, 51] as in the pathogenesis of HF through their ability to regulate the expression levels of genes that govern the process of adaptive and maladaptive CR [52]. Besides, miRNAs can mediate the beneficial effects of ET in heart in both physiological and pathological CR well as in cardiac disorders [2, 53, 54].

Therefore, in the below section we will highlight the role of miRNAs in both physiological CR well as diseased heart. In another section, we will highlight what is known until now about the effects of ET in the regulation of miRNA expression in heart in these different scenarios.

4.2 miRNAs in cardiac remodeling and heart diseases

The miRNAs may be essential regulators in the development and normal physiology. Dysregulation of miRNA expression has been found to correlate with CR and disease in both human and mouse [55, 56]. Therefore, identification of a signature of miRNAs is an essential and promising prerequisite for the study of the biological functions of this class of molecules in the cardiovascular system [57, 58].

A single miRNA can be highly expressed in one tissue and may not have, or have a low expression in another [58]. In this sense, miRNA-1, -133, -206, and -208 are muscle specific and are primarily expressed in the cardiac and skeletal muscle [57, 59], with miRNA-208 being solely expressed in heart [57]. During cardiac development, the expression of miRNA-1 increases, which promotes a decrease in levels of the heart and neural crest derivatives expressed 2 levels (Hand2), a gene-target, reaching it at the levels found in mature cardiac myocytes [57]. miRNA-1 also promotes myogenesis, targeting the histone deacetylase 4 (HDAC4). The miRNA-1 family represents more than 40% of all the miRNAs expressed in the heart [60], which consists of a subfamily composed of miRNA-1-1, miRNA-1-2 and miRNA-206.

The miRNA-133 family consists of the subfamily miRNA-133a-1, 133a-2, and 133b. miRNA-133, like miRNA-1, also plays an important role in cardiac development. In fact, the deletion of miRNA-133 results in severe cardiac malformations, together with embryonic and post-natal lethality, as a consequence of an insufficient number of cardioblasts [57].

The miRNA-208a, -208b, and -499 are specifically encoded by introns of Myh6 (α-MHC), Myh7 (β-MHC), and Myh7b, respectively. In mice, Myh7/miRNA-208b is expressed in the embryonic heart and Myh6/miR-208a is expressed predominantly in the postnatal stages. Myh7 is re-expressed in the adult's heart only after cardiac stress. Myh7b/miRNA-499 is expressed in the embryonic heart and in the adult heart [61]. However, the targeted deletion of miRNA-208a in mice results in only a mild cardiac phenotype, such as the ectopic expression of rapid muscle markers of the skeleton and the impaired ability to respond to postnatal stress. Therefore, this scenario suggests a not so essential role of these three miRNAs during cardiac development [51, 62].

Considering the importance of miRNAs even in heart development, it is understandable that dysregulation in its expression is implicated in a variety of pathological disorders, including diabetes, various cancers, and CVD [51]. Pathological CH, unlike the physiological CH promoted by ET, is a common response to various cardiac disorders, which includes not only hypertension, but also cardiac ischemic disease, valvular diseases, and endocrine disorders, and is therefore, considered the major determinant of mortality and morbidity in CVD [58, 63]. Currently, existent literature suggest that mainly miRNA-1, -18b, -21, -133, -195, and -208 play important roles in modulating CH [52].

In this sense, Carè et al. [64] found a decrease in expression of both miRNA-133 and miRNA-1 in three different animal models of CH, including hypertrophy induced by pressure overload, hypertrophy induced by Akt overexpression, and adaptive CH in mice and humans [64]. However, *in vitro* overexpression of these miRNAs inhibited hypertrophy, while *in vivo* inhibition of miRNA-133 promoted a pronounced and sustained CH. In this same study, three new specific targets for miRNA-133 have been identified, which include RhoA and CH regulatory GDP-GTP exchange protein; the Cdc42, a signal transduction kinase implicated in hypertrophy; and Nelf-A/WHSC2, a nuclear factor involved in cardiogenesis [64]. miRNA-1, in turn, controls cardiac myocyte growth by negatively regulating the expression of calcium-calmodulin signaling components, Mef2a and Gata4, which are key transcription factors that mediate calcium-dependent changes in gene expression [65].

Cheng et al. [55] showed that miRNAs are aberrantly expressed in culture of neonatal cardiomyocytes stimulated with angiotensin II or phenylephrine, and that miRNA-21 knockdown exerts a negative effect on cardiomyocyte hypertrophy *in vitro*. The mitogen-activated protein kinase (MAPK), FasLigand (FasL), and homolog of deleted phosphatase on chromosome 10 (PTEN) are targets that have already been validated for miRNA-21, which stimulate the Akt pathway [66, 67]. In addition, there is evidence that depletion of miRNA-21 exerts anti-apoptotic and pro-proliferative effects on rat proliferative vascular smooth muscle cells (VSMCs) [68]. On the other hand, overexpression of miRNA-1 (which was aberrantly down-regulated in induced-hypertrophy model in mouse), prevented the hypertrophic growth of cardiac myocytes that were stimulated by endothelin-1 [67].

Cardiac fibrosis is an important contributor for the development of cardiac dysfunction in diverse pathological conditions, such as myocardial infarction (MI), ischemic, dilated, hypertrophic cardiomyopathies, and HF and can be defined as an inappropriate accumulation of extracellular matrix proteins in the heart [69]. The exacerbated expression of extracellular matrix proteins is determinant in the differentiation between physiological and pathological CH [2]. In this sense, the negative regulation of miRNA-29 with anti-miRNAs *in vitro* and *in vivo* in a MI model induces the expression of collagens, whereas miRNA-29 overexpression in fibroblasts reduces this expression, suggesting this miRNA as a regulator of cardiac fibrosis [46]. Currently, the miRNA-29 family has been validated as a regulator of the gene expression of collagen I, III, fibrillin I, and elastin and is involved in pathological CH and the HF process [46, 51, 66]. In addition, decreased expression of the miRNA-29 family, miRNA-24, and miRNA-320 occurs after MI, and the decrease in miRNA-29 expression promotes fibrosis and scar formation in the heart [51].

Through microarray analysis, van Rooij et al. [56] related the miRNAs found both in response to hypertrophy and HF and identified more than 12 miRNAs dysregulated in the heart of mice. Many of these miRNAs were similarly regulated in failing human hearts. Interestingly, overexpression of miRNA-23a, miRNA-23b, miRNA-24, miRNA-195, or miRNA-214 induced hypertrophic growth in cardiomyocyte culture, whereas overexpression of miRNA-150 or miRNA-181b decreased cell size of the cardiomyocyte. The miRNA-23a knockdown attenuated hypertrophy [56] and the overexpression of miRNA-195 *in vivo* alone (transgenic mouse), was sufficient to lead to hypertrophy and HF [56]. The miRNA-23a is transcriptionally regulated by the nuclear factor of activated T cells, cytoplasmic 3 (NFATc3), which mediates the signaling pathway of cardiac stress. Thus, miRNA-23a overexpression can triggers a hypertrophic response by means of its target-gene muRF1 (muscle specific ring finger protein 1), an anti-hypertrophic protein, while negative regulation of miRNA-23a abolishes CH induced by isoproterenol in mice [70]. Other miRNAs exhibited pro-hypertrophic profile, including miRNA-208, -21, -18b, -27, and -9 [56].

Transgenic overexpression of miRNA-208a in the adult heart of mice is enough to induce hypertrophic growth, resulting in pronounced repression of regulatory target-genes, such as THRAP1 and myostatin (negative regulators of muscle growth and hypertrophy). In contrast, the deletion of cardiac miRNA-208a protects these animals from CH in response to hemodynamic cardiac stress by canceling the re-expression of fetal gene β-MHC [71]. In addition, therapeutic inhibition of miRNA-208 (by subcutaneous delivery of anti-miRNA-208a) reduces deleterious CR, improves cardiac function, and survival during HF in hypertensive rats [72]. Curiously, inhibition of cardiac-specific miR-208a may have therapeutic usefulness in a variety of metabolic disorders, such as obesity and type-2 diabetes in the setting of cardiac dysfunction by targeting and increasing MED13 in heart [73].

miRNA-199a is predominantly expressed in cardiomyocytes, where it maintains cell size and may play a role in the regulation of CH [56]. miRNA-199 is downregulated under hypoxic conditions in cardiac myocytes (SIRT1) and hypoxia-1α inducible factor (HIF-α), an essential transcription factor for induction of the gene network of response to hypoxia. Thus, miRNA-24, miRNA-29, miRNA-199, and miRNA-320 represent potential therapeutic targets for ischemic injury in cardiac myocytes [51].

An overexpression of miRNA-22 was sufficient to induce pathological CH [74], showing itself as an essential miRNA for this process. On the other hand, the silencing of miRNA-22 in the heart of mice abolished CH and remodeling in response to two independent stressors: the infusion of isoproterenol and the activated calcineurin transgene. In addition, SIRT1 and HDAC4 were validated as target genes for miRNA-22 in the heart [74]. miRNA-155 is also required for the development of CH in response to stress. In fact, the knockout of miRNA-155 in mice suppresses pathological CR, in addition to preventing the progression of HF [75].

The miRNA-34 family (miRNA-34a, -34b, and -34c) expression levels are elevated in both hearts of mice with cardiac stress and in aging process, as well as in patients with HF [76, 77]. In this way, inhibition with antimiR-34a/antimiR-34 has emerged as a promising therapeutic strategy [78]. miRNA-34a appears to be an essential regulator in cardiac repair and regeneration after MI in the heart. In fact, in a study by Yang et al., miRNA-34a mimic delivery limited cardiomyocyte proliferation and subsequent MI recovery in mice, whereas antimiR-34a treatment improved remodeling post infarction. Furthermore, in isolated cardiomyocytes, miRNA-34a directly regulated cell cycle activity and death, via modulation of its targets, which include Bcl2, Cyclin D1, and SIRT1 [79]. Thus, at first, modulation of miRNA-34a appears to be important in the process of repairing adult myocardium after injury. The study conducted by Baker et al. showed that miRNA-34a is induced by oxidative stress (OE) in patients with chronic obstructive pulmonary disease, reducing the gene expression of SIRT1 and SIRT6 in bronchial epithelial cells [80]. Thus, miRNA-34a antigens may have therapeutic potential in several pathological conditions, including CH.

In addition, Bernardo et al. [76] showed that only the inhibition of the whole miRNA-34 family, but not the inhibition of miRNA-34 alone, improves cardiac function in mice with preexisting hypertrophy induced by TAC and also attenuates the pathological remodeling of the LV after MI [76]. In this same work, four targets for miRNA-34 (vinculin, Sema4b, Pofut1, and Bcl6) were validated, which were positively regulated and associated with the cardiac protection found in both TAC and/or MI mice treated with the antimiR-34 LNA. Thus, the authors believe that, since LNA-based therapies have already entered clinical trials, inhibition of the entire miRNA-34 family has great therapeutic potential.

Therefore, it is clear that miRNAs play pivotal roles in cardiovascular development and disease [51, 56]. Recent evidences show that miRNAs are dynamically regulated by exercise [1, 54, 81]. Thus, the elucidation of the relationship between ET and miRNAs in heart in physiological and pathological (i.e., hypertension, HF, spinal cord injury, etc.) conditions is fundamental to understand how exercise regulates the cardiovascular system at the molecular level, which may be promising for the development of new drugs [2].

4.3 miRNAs regulation by exercise training in cardiac remodeling and heart diseases

The miRNA-1 and miRNA-133 are decreased in both aerobic treadmill-induced physiological hypertrophy and pressure overload-induced pathologic hypertrophy [64]. Additionally, swimming training is recognized for its efficiency in inducing myocardial hypertrophy and a significant increase in LV end-diastolic volume in rats [82]. In study conducted by Soci et al. [66], female rats were submitted to two swimming protocols (one moderate and other with high volume training to mimic an active individual and an elite athlete, respectively) for 10 weeks and developed physiological CH in a proportional way to the training volume. It was associated to a down regulation of miRNAs-1, -133a, and -133b in LV of trained rats also compared to sedentary group, which is inversely proportional to the physiological CH [66]. These results suggest that these myomiRs can be regulated by exercise, regardless of type (running or swimming) or volume (moderate and high) training, maintaining a similar expression profile [1]. Besides, Ma et al. [49] found that CH in rats induced by swimming exercise correlates with the modulation of other miRNAs, including miRNA-21, miRNA-124, miRNA-144, and miRNA-145 [49]. These miRNAs have as validated targets Pik3a, PTEN, and TSC2, which are negative regulators of the PI3K/AKT/mTOR signaling pathway. Importantly, Liu et al. [5] showed that the inhibition of miRNA-222 *in vivo* completely blocked cardiac and cardiomyocyte growth in response to exercise while reducing markers of cardiomyocyte proliferation. Indeed, the authors revealed that miRNA-222 delivery was sufficient to protect the heart against dysfunction and adverse remodeling after ischemic injury [5].

In the same work cited before, Soci et al. [66] found an increase in miRNA-29c expression in LV in response to swimming training, which was inversely correlated with a decrease in both collagen I and III expression as well as in OH-proline concentration, being essential for the improvement of LV compliance observed in rats [66]. These findings exhibit an important integrative and regulatory role for miRNAs in the fibrotic response of trained heart.

Other work from our group showed that swimming training also causes an increase in miRNA-126 in the heart of rats, targeting and decreasing in expression of Spred-1, which contributes to angiogenesis in this tissue [83]. Although Pi3kr2 is also a well-validated target for miRNA-126, it was not altered by the swimming protocol. Physiological CH also involves the regulation of some miRNAs related to increased expression of angiotensin receptor-1 (AT1R), regardless of the participation of Ang-II. In parallel, increased angiotensin-converting enzyme 2 (ACE2), angiotensin (1–7), and Ang-II receptor in heart, suggest that miRNAs are involved in the regulation of the non-classical renin-angiotensin-aldosterone system (RAAS), counteracting the classic cardiac RAAS in CH [2, 54]. Fernandes et al. [54] evaluated that swimming alter the expression of cardiac miRNAs targeting the components of the RAAS, which were associated with the development of ventricular CH in rats [54]. In fact, swimming training increased the expression of miRNAs-27a and 27b by targeting the angiotensin-converting enzyme (ACE) and decreased miRNA-143 expression by targeting ACE2 in the heart of healthy rats.

Differently from which is found in pathological CH [61], high volume ET decreases the expression of cardiac miRNA-208a in healthy, induces the upregulation of targets such as THRAP-1, Purβ, and Sox6, and improves the balance between βMHC and αMHC gene expression [2, 84]. In addition, in recent study, Fernandes et al. [85] showed that aerobic ET prevents weight gain and pathological CH in obese Zucker rats, via increase of cardiac MED13 by regulation of miRNA-208. Thus, miRNA-208 represents a potential potent therapeutic target for modulation of cardiac function and remodeling during progression of heart disease.

In addition, the study conducted by Fernandes et al. [86] showed that swimming training reestablishes the peripheral levels of the -16, -21, and -126 miRNAs associated with revascularization in hypertension. This alteration was accompanied by normalization of vascular endothelial growth factor (VEGF), endothelial nitric oxide synthase (eNOS), and phosphatidylinositol-3 kinase regulatory subunit (PI3KR2) in parallel to a normalization of pro-apoptotic (Bad) and anti-apoptotic mediators (Bcl-2, Bcl-x, and p-Badser112: Bad ratio) [86].

Swimming ET restores the miRNA-29a and miRNA-29c levels of the heart and prevents the deposition of type I and III collagen in the border zone as well as in remote regions of the infarcted myocardium of rats, which can contribute to the improvement of the ventricular function induced by the aerobic training [87]. Cardiac fibrosis, which impairs cardiac contractility, is a major aspect of the remodeling process after MI. Impaired cardiomyocyte contractility and calcium transient are hallmarks of LV contractile dysfunction [88]. In this sense, Melo et al. [81] showed that moderate aerobic ET also restores cardiac Ca^{2+} transient after MI, regulating miRNA-1 and miRNA-214 levels [81]. Additionally, resistance training has been shown to induce CH, with improved cardiac function of isolated cardiomyocytes, which is partially explained by the decrease of miRNA-214 and the increase of gene SERCA2a expression [81]. There are few studies that demonstrate the protective role of resistance training in heart through changes in miRNA expression pattern. Thus, there is a potential role of these miRNAs in promoting cardioprotective effects on CR.

5. Vascular, exercise, and miRNAs

The vascular disease in its various forms is one of the leading causes of death worldwide, and in accordance with the projection of the World Health Organization, this number only increases in the next 30 years [89]. ET prevents the progression of vascular diseases and reduces cardiovascular morbidity and mortality. ET also ameliorates vascular changes including endothelial dysfunction and arterial remodeling and stiffness, usually present in type 2 diabetes, obesity, hypertension, and metabolic syndrome [90].

Although there are major advances in the surgical area, the great challenge that anyone who presents a framework of CVD remains the prevention of this disease. Currently, there are some ways to alleviate this disease such as the use of beta-blockers, ACE inhibitors, statins; all these strategies are for mitigate the risk factors such as hypertension and dyslipidemia. Already in more severe cases, surgical procedures such as coronary artery bypass grafting, angioplasty (in severe cases of atherosclerosis), and others are performed [91–93].

Despite these strategies of mitigate the risk factors, this is still not enough since the rate of individuals affected and morbimortality has been increasing in relation to this class of diseases. So, the search for new therapies and approaches to prevent the onset and reduce the progression of vascular disease is desperately needed. Therefore, the processes time consuming since the mechanisms involved in this disease are not yet fully understood.

Atherosclerosis is the main cause of CVD. This is a chronic inflammatory disease that affects the blood vessels and arteries of medium or large size, usually in the areas of bifurcation, due to the non-laminar flow, leading to an endothelial dysfunction due to chronic exposure to pro-inflammatory molecules, and also to an inhibition of anti-inflammatory molecules, such as NO, this causes a narrowing of the arterial lumen due to atheroma plaque, impeding blood flow [94, 95].

In this disease, the blood flow changes from laminar to turbulent, due to the narrowing of the vascular lumen, the endothelial cells (EC) respond to this flow, that is, to the shear stress, this response is due to mechanosensors genes present on the cell surface and also the activation of the adhesion molecules and other inflammatory cytokines that are activated to alleviate the excess of cholesterol low-density lipoprotein (LDL) that is encompassed in the vascular intima [96–98].

Another important factor that has been widely studied in vascular diseases and in atherosclerosis are the miRNAs, which can regulate the cellular phenotype of both VSMCs and EC, such as the proliferation, migration, and cell growth, and can also regulate the handling of lipids and inflammatory molecules [99–101].

5.1 miRNA-126

miRNA-126, for example, can regulate both inflammation and angiogenesis (the process of new vessel formation). This miRNA is basically expressed in platelets and EC, which in turn play an important role in the blood circulation, as well, regulating vascular tone, recruitment of other cells, and even vascular homeostasis [102, 103], and the transcription of this miRNA is regulated by the transcription factor Ets1, which is known for their role in specific gene expression in EC [104]. Harris et al.

[104] demonstrated that miRNA-126 negatively regulates the vascular cell adhesion molecule (VCAM-1). This protein mediates the adhesion of lymphocytes, monocytes, eosinophils, and basophils of the vascular endothelium, thus interfering with the adhesion of leukocytes induced by tumor necrosis factor (TNF) in the endothelium, which may decrease leukocyte infiltration in the vascular wall and also in the interaction of leukocytes with EC [104].

This miRNA can influence the reduction of apoptotic cells contained in the vascular thrombus, softening the inflammatory reaction characteristic of atherosclerosis; so the miRNA-126 has an anti-atherosclerotic effect, which is also related to the G 16 protein signaling regulator, which results in recruitment of circulating progenitor cells, leading to stabilization of atherosclerotic plaque [105, 106]. Studies show that on deletion of this miRNA, vascular integrity decreases; in the case of infarcted individuals, a defect in cardiac neovascularization occurs [107].

This miRNA may be pro-angiogenic because of its binding to the PI3K pathway, which is important in regulating mitogenesis, cell differentiation, and insulin-stimulated glucose transport, and mitogen-activated protein kinases (MAPKs) that respond to extracellular stimuli and regulate various cellular activities such as gene expression, mitosis, differentiation, cell survival, and apoptosis. PI3K pathway promotes the suppression of the sprouty-related EVH1 domain containing protein-1 protein (Spred-1) that generally inhibits the signaling pathway [108–111].

Recent studies showed that the miRNA-126 levels in the circulatory system increase to varying degrees after different ET protocols [1]. It is already known that regular aerobic exercise has a positive effect on arterial endothelial function, mainly by increasing circulating NO and reducing endothelin-1. Aerobic exercise can also improve vascular endothelial and mitochondrial function to protect blood vessels, thus reducing the incidence of CVD [83, 112, 113].

Recent study demonstrated that the plasma levels of miRNA-126 were increased in obese subjects after an acute aerobic exercise session [114] Already noted that improvement of vascular endothelial function in adolescents after 6 weeks of aerobic exercise combined with dietary control may be related to changes in serum level miRNA-126, this improved peripheral vasodilator capacity, indirectly reflecting the improvement of vascular endothelial function in obese adolescents [115]. This increase of serum levels of the miRNA-126 through ET represents the beneficial aspects for obese individuals and associated diseases such as CVD. This may be related to the increase of NO provided by physical training, which is already well established in the literature, that aerobic exercise increases the expression and phosphorylation of endothelial nitric oxide synthase (eNOS) and with that the levels of oxidative stress decreased, leading to an increase in the use of NO and an increase in vascular function [90, 102, 116–118].

5.2 miRNA-21

miRNA-21 regulates cell proliferation through your gene target PTEN, which is a protein that acts as phosphatase and operates primarily on dephosphorylation in the PI3K/Akt signaling pathway. PTEN suppresses Akt signaling, which decreases the activity of eNOS and PTEN also inhibits the expression of VCAM-1 in EC stimulated by TNF-α [119]. A pathological condition that increases the expression of the miRNA-21 is shear stress in your state turbulent flow [120]. miRNA-21 also contributes to phenotypic change in EC, which is induced by transforming growth factor β (TGF-β) [121]. These data indicate that miRNA-21 modulates vascular homeostasis through PTEN and Akt.

Although miRNA-21 has been involved in promoting the proliferation of VSMCs in response to a series of models of vascular mechanical injury, its role remains to be defined in the formation of atherosclerotic lesion [68]. miRNA-21 also mediates VSMC differentiation from the contractile phenotype to the synthetic through TGF-β and BMP signaling in a single posttranscriptional processing step, implying that Smad proteins can control the maturation of miRNAs [122].

It was demonstrated by [86] that aerobic ET has a beneficial effect on the regression of arterial hypertension, restoring the normal expression of miRNA-16, -21, and -126 of skeletal muscle microcirculation, normalization of levels of VEGF, eNOS, and PI3KR2, as well as proapoptotic (bad) and antiapoptotic mediators (Bcl-2, Bcl-x, and p-Badser112: Bad ratio), indicating that the balance between angiogenesis and apoptotic factors that can prevent microvascular abnormalities in rats hypertensive, demonstrating once again the protective effect of aerobic physical training.

5.3 miRNA-221/222

The miRNA-221/222 is expressed in EC, VSMCs, and hematopoietic cells. It has been shown that the endothelial progenitor cells of patients with coronary artery disease have an increased expression of these miRNAs, these cells have a factor important in endothelial integrity and vascular repair [123]. One of the target genes of these miRNAs is signal transducer and activator of transcription 5A (Stat5A), a transcription factor involved in the regulation of cell proliferation and migration, it has been identified that Stat5A is negatively regulated by miR-222 and it has been observed in EC of atherosclerotic lesions that Stat5A is low expression, so it is possible that Stat5A may be a new therapeutic target for atherosclerosis [103, 123, 124].

The study by Liu et al. [123] also demonstrated that miRNA-221/222 has opposite biological functions depending on the cell; the study observed that inhibition of miRNA-221/222 is able to increase reendothelization, whereas in VSMC neointimal formation was significantly reduced. Thus, miRNA-221/222 could be an ideal therapeutic target in patients with atherosclerosis, angioplasty, stent implantation, or myocardial revascularization surgery [123].

5.4 miRNA-146a

The increased expression of the miRNA-146a has been observed in vascular injury after balloon angioplasty. This miRNA can mediate cell-to-cell communication between VSMCs and CE and also other cell types in the vessel wall [125]. A study by Sun et al. [126] attested that transcription of miRNA-146a is regulated by the transcriptional factors KLF-4 (Krüppel-like factor 4) and KLF-5 (Krüppel-like factor 5) and they play an important role in promoting proliferation of VSMCs (*in vitro*) and when the carotid artery is injured with a balloon catheter there occurs a hyperplasia of the neointima (*in vivo*) [126]. Ji et al. [68] also found an increase in miRNA-146a in injured arteries, resulting in a remarkable formation of neointima. In this study, the expression of several miRNAs in the wall of the vessel after angioplasty was observed for the first time [68]. In the study by Chen et al. [127], several miRNAs have been shown to play roles such as proliferation, cell differentiation, and apoptosis in the inflammatory responses of monocytes and macrophages with stimulation of oxidized LDL that are primordial in the development of atherosclerosis [127].

Cao et al. [128] observed that levels of miRNA-146a and miRNA-21 were increased in atherosclerotic plaques and that these miRNAs suppress the expression of Notch2 (neurogenic locus notch homolog protein) which is a membrane protein, and Jag1 (Jagged1) is a cell surface protein [128]. These provide important negative feedback on the proliferation of VSMCs in response to vascular injury, so inhibitions of these miRNAs may present a new strategy for the prevention of fibroproliferative vascular diseases.

5.5 miRNAs-143/145

The miRNAs-143/145 have an important role in the differentiation of VSMC as well as miRNA-146a, with a critical role in cell phenotype change [129, 130]. Boettger et al. [130] demonstrated that the detection of miRNA-143 and miRNA-145 altered the VSMC phenotype from contractile (physiological) to synthetic (pathological), in which this led to neointimal formation in elderly mice [130]. Another study highlighted the suppression of these miRNAs in human aneurysms [131]. Several targets of miRNA-143 and miRNA-145 have been reported in which they regulate the functions of VSMCs, for example, the transcription factors KLF-4 (Krüppel-like factor 4), myocardin, Elk-1 (ELK1, ETS member of the family of oncogenes), and KLF5. In addition, the ACE has been shown to contribute to vascular tonus dysregulation and to reduce blood pressure in mice with high expression of miRNA-143/145 [129, 130]. Thus, these miRNAs through their targets are antiatherosclerotic and have the function of regulating the phenotype of the VSMCs and inhibit the formation of neointima.

5.6 Others miRNAs

Other miRNAs are also involved with EC and VSMCs like miRNA-27b and let-7. Kuehbacher et al. [103] demonstrated that inhibition of miRNA-27b and let-7f in the EC significantly reduced the onset of the endothelium, therefore when these miRNAs are with increased expression, it is suggested that it has a pro-angiogenic action [103]. It has also been observed that shear stress increases the expression of miRNA-27b supporting an activity endothelial protection [132]. Other miRNA vascular protector is the miRNA-296 which regulates directly the substrate of tyrosine kinase regulated by the hepatocyte growth factor (HGS), a protein responsible for the degradation of pro-angiogenic receptors VEGFR2 and PDGFRβ, promoting angiogenesis [133, 134]. The miRNA-130a was also identified as angiogenic, targeting the homeobox anti-angiogenic GAX (homeobox-stop-growth), and HoxA5 proteins. It has been observed that miRNA-130a antagonizes the inhibitory effects of GAX on the proliferation, migration, and formation of EC [135].

That said, miRNAs are important regulators of vascular cell functions and contribute to many vascular diseases, such as coronary artery disease and, in general, atherosclerosis. It is supposed that the human genome encodes several miRNA genes, but only a few vascular-specific miRNAs have been identified so far. In addition, there is only some information about its specific functions of the cell type and target genes. Many studies have revealed an aberrant miRNA expression profile under pathological conditions. Thus, miRNAs can be used as new biomarkers and, on the other hand, may represent powerful new therapeutic targets. Several miRNA-based therapeutic strategies are used to modify miRNA expression as therapeutic approaches; one of miRNA modulators is physical training, as already mentioned in miRNAs-126 and 21.

6. Conclusions

ET, especially aerobic exercise, plays a protective role in the cardiovascular system and has been used as a non-pharmacological therapeutic tool for the prevention and treatment of CVD. Although much knowledge already exists, there is insufficient evidence to establish a direct link between epigenetic modulations and changes, caused by exercise, in the heart and blood vessels [42]. There is currently evidence that the protective effects of aerobic ET involve changes in pattern expression of specific miRNAs by positively regulating cardiovascular remodeling. In fact, miRNAs play important roles in the regulation of distinct processes in mammals, and although they exhibit limited complementarity with their target RNAs, it is still sufficient for them to regulate various physiological processes, including cell proliferation, differentiation, migration, angiogenesis, apoptosis, tissue development, and remodeling [14, 136, 137]. However, new studies to understand the pathways and mechanisms in which the ET acts on the miRNAs are required. Thus, studies with miRNAs hope to broaden the perspective of its use in the correction of pathological processes from miRNAs therapy, contributing as a therapeutic intervention in cardiovascular damages as well as for the survival and quality of life of patients.

Acknowledgements

The researchers were supported by Sao Paulo Research Foundation (FAPESP: 2016/26156-5, 2015/22814-5 and 2015/17275-8), USP/PRP-NAPmiR, National Council for Scientific and Technological Development (CNPq: 307591/2009-3 and 159827/2011-6), and Coordination for the Improvement of Higher Education Personnel (CAPES-Proex).

Conflict of interest

There are no conflicts of interest.

Notes/Thanks/Other declarations

Thanks to the support of data basis Smart Servier Medical Art in the design of the figures by authors (https://smart.servier.com/image-set-download/).

Author details

Noemy Pereira, Camila Gatto, Edilamar Menezes de Oliveira and Tiago Fernandes*
Laboratory of Biochemistry and Molecular Biology of the Exercise, Physical Education and Sports School, University of Sao Paulo, Sao Paulo, Brazil

*Address all correspondence to: tifernandes@usp.br

References

[1] Fernandes T, Baraúna VG, Negrão CE, Phillips MI, Oliveira EM. Aerobic exercise training promotes physiological cardiac remodeling involving a set of microRNAs. American Journal of Physiology. Heart and Circulatory Physiology. 2015;**309**(4):H543-H552

[2] Das NVJ. Exercise training in hypertension: Role of microRNAs. World Journal of Cardiology. 2014;**6**(8):713

[3] Liu X, Platt C, Rosenzweig A. The role of microRNAs in the cardiac response to exercise. Cold Spring Harbor Perspectives in Medicine. 2017;**7**(12):19

[4] O'Connor CM, Whellan DJ, Lee KL, Keteyian SJ, Cooper LS, Ellis SJ, et al. Efficacy and safety of exercise training in patients with chronic heart failure. Journal of the American Medical Association. 2009;**301**(14):1439

[5] Liu X, Xiao J, Zhu H, Wei X, Platt C, Damilano F, et al. miR-222 is necessary for exercise-induced cardiac growth and protects against pathological cardiac remodeling. Cell Metabolism. 2015;**21**(4):584-595

[6] Boon RA. Non-coding RNAs in cardiovascular health and disease. Non-coding RNA Research. 2018;**3**(3):99

[7] Palazzo AF, Lee ES. Non-coding RNA: What is functional and what is junk? Frontiers in Genetics. 2015;**6**:2. Available from: http://journal.frontiersin.org/article/10.3389/fgene.2015.00002/abstract

[8] Rotini A, Martínez-Sarrà E, Pozzo E, Sampaolesi M. Interactions between microRNAs and long non-coding RNAs in cardiac development and repair. Pharmacological Research. 2018;**127**:58-66. Available from: https://linkinghub.elsevier.com/retrieve/pii/S1043661817303572

[9] Bartolomei MS, Zemel S, Tilghman SM. Parental imprinting of the mouse H19 gene. Nature. 1991;**351**(6322):153-155. Available from: http://www.nature.com/articles/351153a0

[10] Poller W, Dimmeler S, Heymans S, Zeller T, Haas J, Karakas M, et al. Non-coding RNAs in cardiovascular diseases: Diagnostic and therapeutic perspectives. European Heart Journal. 2018;**39**(29):2704-2716. Available from: https://academic.oup.com/eurheartj/article/39/29/2704/3738809

[11] Kumarswamy R, Thum T. Non-coding RNAs in cardiac remodeling and heart failure. Circulation Research. 2013;**113**(6):676-689. DOI: 10.1161/CIRCRESAHA.113.300226

[12] Di Mauro V, Barandalla-Sobrados M, Catalucci D. The noncoding-RNA landscape in cardiovascular health and disease. Non-coding RNA Research. 2018;**3**(1):12-19. Available from: https://linkinghub.elsevier.com/retrieve/pii/S2468054017300483

[13] Xu J, Zhao J, Evan G, Xiao C, Cheng Y, Xiao J. Circulating microRNAs: Novel biomarkers for cardiovascular diseases. Journal of Molecular Medicine. 2012;**90**(8):865-875. Available from: http://link.springer.com/10.1007/s00109-011-0840-5

[14] Gomes CPC, de Gonzalo-Calvo D, Toro R, Fernandes T, Theisen D, Wang D-Z, et al. Non-coding RNAs and exercise: Pathophysiological role and clinical application in the cardiovascular system. Clinical Science. 2018;**132**(9):925-942. DOI: 10.1042/CS20171463

[15] Bartel DP. microRNAs. Cell. 2004;**116**(2):281-297. Available from: http://linkinghub.elsevier.com/retrieve/pii/S0092867404000455

[16] Ha M, Kim VN. Regulation of microRNA biogenesis. Nature Reviews. Molecular Cell Biology. 2014;**15**(8):509-524. Available from: http://www.nature.com/articles/nrm3838

[17] Condorelli G, Latronico MVG, Cavarretta E. microRNAs in cardiovascular diseases. Journal of the American College of Cardiology. 2014;**63**(21):2177-2187. Available from: https://linkinghub.elsevier.com/retrieve/pii/S0735109714011085

[18] Moretti F, Thermann R, Hentze MW. Mechanism of translational regulation by miR-2 from sites in the 5′ untranslated region or the open reading frame. RNA. 2010;**16**(12):2493-2502. DOI: 10.1261/rna.2384610

[19] Lee RC, Feinbaum RL, Ambros V. The *C. elegans* heterochronic gene lin-4 encodes small RNAs with antisense complementarity to lin-14. Cell. 1993;**75**(5):843-854. Available from: http://linkinghub.elsevier.com/retrieve/pii/009286749390529Y

[20] Lewis BP, Burge CB, Bartel DP. Conserved seed pairing, often flanked by adenosines, indicates that thousands of human genes are MicroRNA targets. Cell. 2005;**120**(1):15-20. Available from: http://linkinghub.elsevier.com/retrieve/pii/S0092867404012607

[21] Gomes CPC, Spencer H, Ford KL, Michel LYM, Baker AH, Emanueli C, et al. The function and therapeutic potential of long non-coding RNAs in cardiovascular development and disease. Molecular Therapy--Nucleic Acids. 2017;**8**:494-507. Available from: https://linkinghub.elsevier.com/retrieve/pii/S2162253117302251

[22] Quinn JJ, Chang HY. Unique features of long non-coding RNA biogenesis and function. Nature Reviews. Genetics. 2016;**17**(1):47-62. Available from: http://www.ncbi.nlm.nih.gov/pubmed/26666209

[23] Ponting CP, Oliver PL, Reik W. Evolution and function so f long noncoding RNAs. Cell. 2005;**136**(4):629-641. Available from: http://www.ncbi.nlm.nih.gov/pubmed/19239885

[24] Liu L, An X, Li Z, Song Y, Li L, Zuo S, et al. The H19 long noncoding RNA is a novel negative regulator of cardiomyocyte hypertrophy. Cardiovascular Research. 2016;**111**(1):56-65. DOI: 10.1093/cvr/cvw078

[25] Han P, Li W, Lin C-H, Yang J, Shang C, Nurnberg ST, et al. A long noncoding RNA protects the heart from pathological hypertrophy. Nature. 2014;**514**(7520):102-106. Available from: http://www.nature.com/articles/nature13596

[26] Wang K, Liu F, Zhou LY, Long B, Yuan SM, Wang Y, et al. The long noncoding RNA CHRF regulates cardiac hypertrophy by targeting miR-489. Circulation Research. 2014;**114**(9):1377-1388

[27] Wang Z, Zhang X-J, Ji Y-X, Zhang P, Deng K-Q, Gong J, et al. The long noncoding RNA Chaer defines an epigenetic checkpoint in cardiac hypertrophy. Nature Medicine. 2016;**22**(10):1131-1139. Available from: http://www.ncbi.nlm.nih.gov/pubmed/27618650

[28] Viereck J, Kumarswamy R, Foinquinos A, Xiao K, Avramopoulos P, Kunz M, et al. Long noncoding RNA Chast promotes cardiac remodeling. Science Translational Medicine. 2016;**8**(326):326ra22. Available from: http://www.ncbi.nlm.nih.gov/pubmed/26888430

[29] Piccoli M-T, Gupta SK, Viereck J, Foinquinos A, Samolovac S, Kramer FL, et al. Inhibition of the cardiac fibroblast-enriched lncRNA Meg3 prevents cardiac fibrosis and diastolic

dysfunction. Circulation Research. 2017;**121**(5):575-583. DOI: 10.1161/CIRCRESAHA.117.310624

[30] Micheletti R, Plaisance I, Abraham BJ, Sarre A, Ting CC, Alexanian M, et al. The long noncoding RNA Wisper controls cardiac fibrosis and remodeling. Science Translational Medicine. 2017;**9**(395):pii: eaai9118

[31] Tao H, Cao W, Yang J-J, Shi K-H, Zhou X, Liu L-P, et al. Long noncoding RNA H19 controls DUSP5/ERK1/2 axis in cardiac fibroblast proliferation and fibrosis. Cardiovascular Pathology. 2016;**25**(5):381-389

[32] Devlin AM, Bottiglieri T, Domann FE, Lentz SR. Tissue-specific changes in H19 methylation and expression in mice with hyperhomocysteinemia. The Journal of Biological Chemistry. 2005;**280**(27):25506-25511. DOI: 10.1074/jbc.M504815200

[33] Kallen AN, Zhou X-B, Xu J, Qiao C, Ma J, Yan L, et al. The imprinted H19LncRNA antagonizes Let-7 microRNAs. Molecular Cell. 2013;**52**(1):101-112. Available from: https://linkinghub.elsevier.com/retrieve/pii/S1097276513006278

[34] Hill JA, Olson EN. Cardiac plasticity. The New England Journal of Medicine. 2008;**358**(13):1370-1380. DOI: 10.1056/NEJMra072139

[35] Porrello ER. microRNAs in cardiac development and regeneration. Clinical Science. 2013;**125**(4):151-166

[36] Van Berlo JH, Maillet M, Molkentin JD. Signaling effectors underlying pathologic growth and remodeling of the heart. Journal of Clinical Investigation. 2013;**123**(1):37-45

[37] Maillet M, Van Berlo JH, Molkentin JD. Molecular basis of physiological heart growth: Fundamental concepts and new players. Nature Reviews. Molecular Cell Biology. 2013;**14**(1):38-48. DOI: 10.1038/nrm3495

[38] Chen C, Ponnusamy M, Liu C, Gao J, Wang K, Li P. MicroRNA as a therapeutic target in cardiac remodeling. BioMed Research International. 2017;**2017**:1278436. DOI: 10.1155/2017/1278436

[39] Bernardo BC, Weeks KL, Pretorius L, McMullen JR. Molecular distinction between physiological and pathological cardiac hypertrophy: Experimental findings and therapeutic strategies. Pharmacology & Therapeutics. 2010;**128**(1):191-227. DOI: 10.1016/j.pharmthera.2010.04.005

[40] Orenes-Piñero E, Montoro-García S, Patel JV, Valdés M, Marín F, Lip GYH. Role of microRNAs in cardiac remodelling: New insights and future perspectives. International Journal of Cardiology. 2013;**167**(5):1651-1659. DOI: 10.1016/j.ijcard.2012.09.120

[41] Burchfield JS, Xie M, Hill JA. Pathological ventricular remodeling: Mechanisms: Part 1 of 2. Circulation. 2013;**128**(4):388-400

[42] Cavalcante PAM, Perilhão MS, Da Silva AA, Serra AJ, Júnior AF, Bocalini DS. Cardiac remodeling and physical exercise: A brief review about concepts and adaptations. International Journal of Sports Science. 2016;**6**(2):52-61

[43] McMullen JR, Jennings GL. Differences between pathological and physiological cardiac hypertrophy: Novel therapeutic strategies to treat heart failure. Clinical and Experimental Pharmacology & Physiology. 2007;**34**(4):255-262

[44] Fernandes T, Soci UPR, Oliveira EM. Eccentric and concentric cardiac hypertrophy induced by exercise training: microRNAs and molecular

determinants. Brazilian Journal of Medical and Biological Research. 2011;**44**(9):836-847

[45] Rivera-Brown AM, Frontera WR. Principles of exercise physiology: Responses to acute exercise and long-term adaptations to training. PM&R: The Journal of Injury, Function, and Rehabilitation. 2012;**4**(11):797-804. DOI: 10.1016/j.pmrj.2012.10.007

[46] van Rooij E, Sutherland LB, Thatcher JE, DiMaio JM, Naseem RH, Marshall WS, et al. Dysregulation of microRNAs after myocardial infarction reveals a role of miR-29 in cardiac fibrosis. Proceedings of the National Academy of Sciences of the United States of America. 2008;**105**(35):13027-13032. Available from: http://www.ncbi.nlm.nih.gov/pubmed/18723672 [Cited: March 15, 2018]

[47] Medeiros A, Oliveira EM, Gianolla R, Casarini DE, Negrão CE, Brum PC. Swimming training increases cardiac vagal activity and induces cardiac hypertrophy in rats. Brazilian Journal of Medical and Biological Research. 2004;**37**(37):1909-1917. Available from: http://www.scielo.br/pdf/bjmbr/v37n12/5413.pdf [Cited: March 15, 2018]

[48] Evangelista FS, Brum PC, Krieger JE, De Medicina F. Duration-controlled swimming exercise training induces cardiac hypertrophy in mice. Brazilian Journal of Medical and Biological Research. 2003;**36**:1751-1759

[49] Ma Z, Qi J, Meng S, Wen B, Zhang J. Swimming exercise training-induced left ventricular hypertrophy involves microRNAs and synergistic regulation of the PI3K/AKT/mTOR signaling pathway. European Journal of Applied Physiology. 2013;**113**(10):2473-2486

[50] Masi LN, Serdan TDA, Levada-Pires AC, Hatanaka E, Silveira LDR, Cury-Boaventura MF, et al. Regulation of gene expression by exercise-related microRNAs. Cellular Physiology and Biochemistry. 2016;**39**(6):2381-2397

[51] Hata A. Functions of microRNAs in cardiovascular biology and disease. Annual Review of Physiology. 2013;**75**(1):69-93. DOI: 10.1146/annurev-physiol-030212-183737

[52] Divakaran V, Mann DL. The emerging role of microRNAs in cardiac remodeling and heart failure. Circulation Research. 2008;**103**(10):1072-1083. DOI: 10.1161/CIRCRESAHA.108.183087

[53] Bei Y, Tao L, Cretoiu D, Cretoiu SM. MiRNAs mediate beneficial effects of exercise in heart. In: Xiao J, editor. Exercise for Cardiovascular Disease Prevention and Treatment. Vol 1000: From Molecular to Clinical, Part 2. 2017. pp. 261-280

[54] Fernandes T, Hashimoto NY, Magalhães FC, Fernandes FB, Casarini DE, Carmona AK, et al. Aerobic exercise training-induced left ventricular hypertrophy involves regulatory microRNAs, decreased angiotensin-converting enzyme-angiotensin II, and synergistic regulation of angiotensin-converting enzyme 2-angiotensin (1-7). Hypertension. 2011;**58**(2):182-189

[55] Cheng Y, Ji R, Yue J, Yang J, Liu X, Chen H, et al. microRNAs are aberrantly expressed in hypertrophic heart: Do they play a pole in cardiac hypertrophy? The American Journal of Pathology. 2007;**170**(6):1831-1840. DOI: 10.2353/ajpath.2007.061170

[56] van Rooij E, Sutherland LB, Liu N, Williams AH, McAnally J, Gerard RD, et al. A signature pattern of stress-responsive microRNAs that can evoke cardiac hypertrophy and heart failure. Proceedings of the National Academy of Sciences. 2006;**103**(48):18255-18260. DOI: 10.1073/pnas.0608791103

[57] Xiao J, Chen Y-H. microRNAs: Novel regulators of the heart. Journal of Thoracic Disease. 2010;**2**(1):43-47. Available from: http://www.pubmedcentral.nih.gov/articlerender.fcgi?artid=3256441&tool=pmcentrez&rendertype=abstract

[58] Zhang C. microRNAs: Role in cardiovascular biology and disease. Clinical Science. 2008;**114**(12):699-706. DOI: 10.1042/CS20070211

[59] Bauersachs J, Thum T. Biogenesis and regulation of cardiovascular microRNAs. Circulation Research. 2011;**109**(3):334-347

[60] Lagos-Quintana M, Rauhut R, Yalcin A, Meyer J, Lendeckel W, Tuschl T. Identification of tissue-specific microRNAs from mouse. Current Biology. 2002;**12**(9):735-739

[61] van Rooij E, Quiat D, Johnson BA, Sutherland LB, Qi X, Richardson JA, et al. A family of microRNAs encoded by myosin genes governs myosin expression and muscle performance. Developmental Cell. 2009;**17**(5):662-673. Available from: http://www.ncbi.nlm.nih.gov/pubmed/19922871 [Cited: January 6, 2019]

[62] van Rooij E, Sutherland LB, Qi X, Richardson JA, Hill J, Olson EN. Control of stress-dependent cardiac growth and gene expression by a MicroRNA. Science. 2007;**316**(5824):575-579. Available from: http://www.ncbi.nlm.nih.gov/pubmed/17379774 [Cited: September 25, 2018]

[63] Drazner MH. The progression of hypertensive heart disease. Circulation. 2011;**123**(3):327-334. DOI: 10.1161/CIRCULATIONAHA.108.845792

[64] Carè A, Catalucci D, Felicetti F, Bonci D, Addario A, Gallo P, et al. MicroRNA-133 controls cardiac hypertrophy. Nature Medicine. 2007;**13**(5):613-618

[65] Ikeda S, He A, Kong SW, Lu J, Bejar R, Bodyak N, et al. MicroRNA-1negatively regulates expression of the hypertrophy-associated calmodulin and Mef2a genes. Molecular and Cellular Biology. 2009;**29**(8):2193-2204. Available from: http://www.ncbi.nlm.nih.gov/pubmed/19188439 [Cited: January 7, 2019]

[66] Soci UPR, Fernandes T, Hashimoto NY, Mota GF, Amadeu MA, Rosa KT, et al. microRNAs 29 are involved in the improvement of ventricular compliance promoted by aerobic exercise training in rats. Physiological Genomics. 2011;**43**(11):665-673

[67] Sayed D, Hong C, Chen IY, Lypowy J, Abdellatif M. microRNAs play an essential role in the development of cardiac hypertrophy. Circulation Research. 2007;**100**(3):416-424

[68] Ji R, Cheng Y, Yue J, Yang J, Liu X, Chen H, et al. MicroRNA expression signature and antisense-mediated depletion reveal an essential role of MicroRNA in vascular neointimal lesion formation. Circulation Research. 2007;**100**(11):1579-1588

[69] Ono K, Kuwabara Y, Han J. microRNAs and cardiovascular diseases. The FEBS Journal. 2011;**278**(10):1619-1633. Available from: http://www.ncbi.nlm.nih.gov/pubmed/21395978 [Cited: January 7, 2019]

[70] Lin Z, Murtaza I, Wang K, Jiao J, Gao J, Li P-F. miR-23a functions downstream of NFATc3 to regulate cardiac hypertrophy. Proceedings of the National Academy of Sciences. 2009;**106**(29):12103-12108. DOI: 10.1073/pnas.0811371106

[71] Callis TE, Pandya K, Hee YS, Tang RH, Tatsuguchi M, Huang ZP, et al. MicroRNA-208a is a regulator of cardiac hypertrophy and conduction in mice. Journal of Clinical Investigation. 2009;**119**(9):2772-2786

[72] Montgomery RL, Hullinger TG, Semus HM, Dickinson BA, Seto AG, Lynch JM, et al. Therapeutic inhibition of miR-208a improves cardiac function and survival during heart failure. Circulation. 2011;**124**(14):1537-1547

[73] Grueter CE, van Rooij E, Johnson BA, DeLeon SM, Sutherland LB, Qi X, et al. A cardiac microRNA governs systemic energy homeostasis by regulation of MED13. Cell. 2012;**149**(3):671-683. Available from: http://www.ncbi.nlm.nih.gov/pubmed/22541436 [Cited: January 7, 2019]

[74] Huang ZP, Chen J, Seok HY, Zhang Z, Kataoka M, Hu X, et al. MicroRNA-22 regulates cardiac hypertrophy and remodeling in response to stress. Circulation Research. 2013;**112**(9):1234-1243

[75] Seok HY, Chen J, Kataoka M, Huang Z-P, Ding J, Yan J, et al. Loss of MicroRNA-155 protects the heart from pathological cardiac hypertrophy. Circulation Research. 2014; **114**(10):1585-1595. Available from: http://www.ncbi.nlm.nih.gov/pubmed/24657879 [Cited: January 7, 2019]

[76] Bernardo BC, Gao X, Winbanks CE, Boey EJH, Keat Y, Kiriazis H. Therapeutic inhibition of the miR-34 family attenuates pathological cardiac remodeling and improves heart function. Proceedings of the National Academy of Sciences of the United States of America. 2012;**109**(43):17615-17620

[77] Boon RA, Iekushi K, Lechner S, Seeger T, Fischer A, Heydt S, et al. MicroRNA-34a regulates cardiac ageing and function. Nature. 2013;**495**(7439):107-110

[78] Ooi JYY, Bernardo BC, Singla S, Patterson NL, Lin RCY, Mcmullen JR. Identification of miR-34 regulatory networks in settings of disease and anti miR-therapy: Implications for treating cardiac pathology and other diseases. RNA Biology. 2017;**14**(5):500-513. DOI: 10.1080/15476286.2016.1181251

[79] Yang Y, Cheng H, Qiu Y, Dupee D, Noonan M, Lin Y, et al. MicroRNA-34aplays a key role in cardiac repair and regeneration following myocardial infarction. Circulation Research. 2015;**117**(5):450-459

[80] Baker JR, Vuppusetty C, Colley T, Papaioannou AI, Fenwick P, Donnelly L, et al. Oxidative stress dependent microRNA-34a activation via PI3Kαreduces the expression of sirtuin-1 andsirtuin-6 in epithelial cells. Scientific Reports. 2016;**21**(6):35871. DOI: 10.1038/ Srep35871

[81] Melo SFS, Barauna VG, Neves VJ, Fernandes T, Lara LDS, Mazzotti DR, et al. Exercise training restores the cardiac microRNA-1 and -214 levels regulating Ca^{2+} handling after myocardial infarction. BMC Cardiovascular Disorders. 2015;**15**:166

[82] Silva GJJ, Bye A, el Azzouzi H, Wisløff U. microRNAs as important regulators of exercise adaptation. Progress in Cardiovascular Diseases. 2017;**60**(1):130-151. DOI: 10.1016/j.pcad.2017.06.003

[83] da Silva ND, Fernandes T, Soci UPR, Monteiro AWA, Phillips MI, DeOliveira EM. Swimming training in rats increases cardiac MicroRNA-126 expression and angiogenesis. Medicine and Science in Sports and Exercise. 2012;**44**(8):1453-1462

[84] Soci UPR, Fernandes T, Barauna VG, Hashimoto NY, de FatimaAlves MG, Rosa KT, et al. Epigenetic control of exercise training-induced cardiac hypertrophy by miR-208. Clinical Science. 2016;**130**(22):2005-2015. DOI: 10.1042/CS20160480

[85] Fernandes T, Barretti DL, Phillips MI, Menezes

Oliveira E. Exercise training prevents obesity-associated disorders: Role of miRNA-208a and MED13. Molecular and Cellular Endocrinology. 2018;**476**:148-154. DOI: 10.1016/j.mce.2018.05.004

[86] Fernandes T, Magalhães F, Roque FR, Phillips MI, Oliveira EM. Exercise training prevents the microvascular rarefaction in hypertension balancing angiogenic and apoptotic factors: Role of microRNAs-16, -21, and -126. Hypertension. 2012;**59**:513-520

[87] Melo SFS, Fernandes T, Baraúna VG, Matos KC, Santos AAS, Tucci PJF, et al. Expression of MicroRNA-29 and collagen in cardiac muscle after swimming training in myocardial-infarcted rats. Cellular Physiology and Biochemistry. 2014;**33**(3):657-669. Available from: http://www.ncbi.nlm.nih.gov/pubmed/24642957 [Cited: March 15, 2018]

[88] Huang Z, Neppl RL, Wang D. microRNAs in cardiac remodeling and disease. Journal of Cardiovascular Translational Research. 2010;**3**(3):212-218

[89] WHO. Global Status Report on Noncommunicable Diseases. Vol. 176. World Health Organization; 2014

[90] Roque FR, Hernanz R, Salaices M, Briones AM. Exercise training and cardiometabolic diseases: Focus on the vascular system. Current Hypertension Reports. 2013;**15**(3):204-214

[91] He X, Zhao M, Bi X, Sun L, Yu X, Zhao M, et al. Novel strategies and underlying protective mechanisms of modulation of vagal activity in cardiovascular diseases. British Journal of Pharmacology. 2015;**172**(23):5489-5500

[92] Barton M. Mechanisms and therapy of atherosclerosis and its clinical complications. Current Opinion in Pharmacology. 2013;**13**(2):149-153

[93] Lundberg JO, Gladwin MT, Weitzberg E. Strategies to increase nitric oxide signalling in cardiovascular disease. Nature Reviews. Drug Discovery. 2015;**14**(9):623-641

[94] Tall AR, Jiang XC, Luo Y, Silver D. 1999 George Lyman Duff memorial lecture: Lipid transfer proteins, HDL metabolism, and atherogenesis. Arteriosclerosis, Thrombosis, and Vascular Biology. 2000;**20**:1185-1188

[95] Barter PJ, Nicholls S, Rye KA, Anantharamaiah GM, Navab M, Fogelman AM. Antiinflammatory properties of HDL. Circulation Research. 2004;**95**:764-772

[96] Dobrin PB, Littooy FN, Endean ED. Mechanical factors predisposing to intimal hyperplasia and medial thickening in autogenous vein grafts. Surgery. 1989;**105**:393-400

[97] Jessup W, Wilson P, Gaus K, Kritharides L. Oxidized lipoproteins and macrophages. Vascular Pharmacology. 2002;**38**:239-248

[98] Mendis S, Puska P, Norrving B. Global Atlas on Cardiovascular Disease Prevention and Control. World Health Organization. Retrieved from: http://whqliboc.who.int/publications/2011/9789241564373_eng.pdf

[99] Tang YL, Zhu W, Cheng M, Chen L, Zhang J, Sun T, et al. Hypoxic preconditioning enhances the benefit of cardiac progenitor cell therapy for treatment of myocardial infarction by inducing CXCR4 expression. Circulation Research. 2009;**104**(10):1209-1216

[100] Moroi M, Zhang L, Yasuda T, Virmani R, Gold HK, Fishman MC, et al. Interaction of genetic deficiency of endothelial nitric oxide, gender,

and pregnancy in vascular response to injury in mice. Journal of Clinical Investigation. 1998;**101**(6):1225-1232

[101] Ross R. The pathogenesis of atherosclerosis: A perspective for the 1990s. Nature. 1993;**362**:801-809

[102] Urbich C, Kuehbacher A, Dimmeler S. Role of microRNAs in vascular diseases, inflammation, and angiogenesis. Cardiovascular Research. 2008;**79**(4):581-588

[103] Kuehbacher A, Urbich C, Zeiher A, Dimmeler S. Role of Dicer and Drosha for endothelial microRNA expression and angiogenesis. Circulation Research. 2007;**101**:59-68

[104] Harris T, Yamakuchi M, Kondo M, Oettgen P, Lowenstein C. Ets-1 and Ets-2 regulate the expression of microRNA-126 in endothelial cells. Arteriosclerosis, Thrombosis, and Vascular Biology. 2010;**30**:1990-1997

[105] Zernecke A, Bidzhekov K, Noels H. Delivery of microRNA-126 by apoptotic bodies induces CXCL12-dependent vascular protection. Science Signaling. 2009;**2**(100):ra81

[106] Weber C, Schober A, Zernecke A. microRNAs in arterial remodelling, inflammation and atherosclerosis. Current Drug Targets. 2010;**11**(8):950-956

[107] Wang S, Aurora A, Johnson B. The endothelial-specific microRNA miR-126 governs vascular integrity and angiogenesis. Developmental Cell. 2008;**15**(2):261-271

[108] Nonami A, Kato R, Taniguchi K, Yoshiga D, Taketomi T, Fukuyama S, et al. Spred-1 negatively regulates interleukin-3-mediated ERK/mitogen-activated protein (MAP) kinase activation in hematopoietic cells. The Journal of Biological Chemistry. 2004;**279**:52543-52551

[109] Taniguchi K, Kohno R, Ayada T, Kato R, Ichiyama K, Morisada T, et al. Spreds are essential for embryonic lymphangiogenesis by regulating vascular endothelial growth factor receptor 3 signaling. Molecular and Cellular Biology. 2007;**27**:4541-4550

[110] Wakioka T, Sasaki A, Kato R, Shouda T, Matsumoto A, Miyoshi K, et al. Spred is a sprouty-related suppressor of ras signalling. Nature. 2001;**412**:647-651

[111] Ueki K, Fruman D, Yballe C, Fasshauer M, Klein J, Asano T, et al. Positive and negative roles of p85 alpha and p85 beta regulatory subunits of phosphoinositide 3-kinase in insulin signaling. The Journal of Biological Chemistry. 2003;**278**:48453-48466

[112] Wu XD, Zeng K, Liu WL, Gao YG, Gong CS, Zhang CX, et al. Effect of aerobic exercise on miRNA-TLR4 signaling in atherosclerosis. International Journal of Sports Medicine. 2014;**35**(4):344-350

[113] Higashi Y, Yoshizumi M. Exercise and endothelial function: Role of endothelium-derived nitric oxide and oxidative stress in healthy subjects and hypertensive patients. Pharmacology & Therapeutics. 2004;**102**(1):87-96

[114] Bao F, Slusher AL, Whitehurst M, Huang CJ. Circulating microRNAs are upregulated following acute aerobic exercise in obese individuals. Physiology & Behavior. 2018;**197**:15-21

[115] Donghui T, Shuang B, Xulong L, Meng Y, Yujing G, Yujie H, et al. Improvement of microvascular endothelial dysfunction induced by exercise and diet is associated with microRNA-126 in obese. Microvascular Research. 2019;**123**:86-91

[116] De Souza CA, Shapiro LF, Clevenger CM, Dinenno FA, Monahan KD, Tanaka H, et al. Regular

aerobic exercise prevents and restore sage-related declines in endothelium-dependent vasodilation in healthy men. Circulation. 2000;**102**(12):1351-1357

[117] Sharma S, Liu J, Wei J, Yuan H, Zhang T, Bishopric NH. Repression of miR-142 by p300 and MAPK is required for survival signalling via gp130 during adaptive hypertrophy. EMBO Molecular Medicine. 2012;**4**(7):617-632

[118] Dyakova EY, Kapilevich LV, Shylko VG, Popov SV, Anfinogenova Y. Physical exercise associated with NO production: Signaling pathways and significance in health and disease. Frontiers in Cell and Development Biology. 2015;**3**:19

[119] Tsoyi K, Jang HJ, Nizamutdinova IT, Park K, Kim YM, Kim HJ, et al. PTEN differentially regulates expressions of ICAM-1 andVCAM-1 through PI3K/Akt/GSK-3β/GATA-6 signaling pathways in TNF-α-activated human endothelial cells. Atherosclerosis. 2010;**213**(1):115-121

[120] Sabatel C, Malvaux L, Bovy N, Deroanne C, Lambert V, Gonzalez M, et al. MicroRNA-21 exhibits antiangiogenic function by targeting RhoB expression in endothelial cells. PLoS One. 2011;**6**(2):169-179

[121] Kumarswamy R, Volkmann I, Jazbutyte V, Dangwal S, Park DH, Thum T. Transforming growth factor-β-induced endothelial-to-mesenchymal transition is partly mediated by MicroRNA-21. Arteriosclerosis, Thrombosis, and Vascular Biology. 2012;**32**(2):361-369

[122] Davis BN, Hilyard AC, Lagna G, Hata A. SMAD proteins control DROSHA-mediated microRNA maturation. Nature. 2008;**454**(7200):56-61

[123] Liu X, Cheng Y, Zhang S, Lin Y, Yang J, Zhang CA. Necessary role of miR-221 and miR-222 in vascular smooth muscle cell proliferation and neointimal hyperplasia. Circulation Research. 2009;**104**(4):476-487

[124] Dentelli P, Rosso A, Orso F, Olgasi C, Taverna D, Brizzi MF. MicroRNA-222 controls neovascularization by regulating signal transducer and activator of transcription 5A expression. Arteriosclerosis, Thrombosis, and Vascular Biology. 2010;**30**(8):1562-1568

[125] Hergenreider E, Heydt S, Tréguer K, Boettger T, Horrevoets AJG, Zeiher AM, et al. Atheroprotective communication between endothelial cells and smooth muscle cells through miRNAs. Nature Cell Biology. 2012;**14**(3):249-256

[126] Sun S, Zheng B, Han M, Fang X, Li H, Miao S, et al. miR-146a and Krüppel-like factor 4 form a feedback loop to participate in vascular smooth muscle cell proliferation. EMBO Reports. 2011;**12**(1):56-62

[127] Chen T, Huang Z, Wang L, Wang Y, Wu F, Meng S, et al. MicroRNA-125a-5p partly regulates the inflammatory response, lipid uptake, and ORP9 expression in oxLDL-stimulated monocyte/macrophages. Cardiovascular Research. 2009;**83**(1):131-139

[128] Cao J, Zhang K, Zheng J, Dong R. MicroRNA-146a and -21 cooperate to regulate vascular smooth muscle cell proliferation via modulation of the notch signaling pathway. Molecular Medicine Reports. 2015;**11**(4):2889-2895

[129] Xin M, Small EM, Sutherland LB, Qi X, McAnally J, Plato CF, et al. microRNAs miR-143 and miR-145 modulate cytoskeletal dynamics and responsiveness of smooth muscle cells to injury. Genes & Development. 2009;**23**(18):2166-2178

[130] Boettger T, Beetz N, Kostin S, Schneider J, Krüger M, Hein L, et al. Acquisition of the contractile phenotype by murine arterial smooth muscle cells depends on the Mir143/145 gene cluster. Journal of Clinical Investigation. 2009;**119**(9):2634-2647

[131] Elia L, Quintavalle M, Zhang J. The knockout of miR-143 and -145 alters smooth muscle cell maintenance and vascular homeostasis in mice: Correlates with human disease. Cell Death and Differentiation. 2009;**16**(12):1590-1598

[132] Wang K-C, Garmire LX, Young A, Nguyen P, Trinh A, Subramaniam S, et al. Role of microRNA-23b in flow-regulation of Rb phosphorylation and endothelial cell growth. Proceedings of the National Academy of Sciences. 2010;**107**(7):3234-3239

[133] Ewan LC, Jopling HM, Jia H, Mittar S, Bagherzadeh A, Howell GJ, et al. Intrinsic tyrosine kinase activity is required for vascular endothelial growth factor receptor 2 ubiquitination, sorting and degradation in endothelial cells. Traffic. 2006;**7**(9):1270-1282

[134] Takata H, Kato M, Denda K, Kitamura NA. Hrs binding protein having a Src homology 3 domain is involved in intracellular degradation of growth factors and their receptors. Genes to Cells. 2000;**5**(1):57-69

[135] Chen Y, Gorski D. Regulation of angiogenesis through a microRNA (miR-130a) that down-regulates antiangiogenic homeobox genes GAX and HOXA5. Blood. 2008;**111**(3):1217-1226

[136] das Neves VJ, Fernandes T, Roque FR, Soci UPR, Melo SFS, de Oliveira EM. Exercise training in hypertension: Role of microRNAs. World Journal of Cardiology. 2014;**6**(8713):27. Available from: http://www.ncbi.nlm.nih.gov/pubmed/25228951 [Cited: September 10, 2018]

[137] Yu P, Wang H, Xie Y, Zhou J, Yao J, Che L. Deregulated cardiac specific microRNAs in postnatal heart growth. BioMed Research International. 2016;**2016**:6241763. DOI: 10.1155/2016/6241763

Section 2

Muscle Disorder: Inflammation-Induced

Chapter 5

Inflammatory Muscle Diseases

Doris Hissako Sumida, Fernando Yamamoto Chiba and Maria Sara de Lima Coutinho Mattera

Abstract

Inflammatory myopathies, also called idiopathic inflammatory myopathy or myositis, are rare conditions characterized by the involvement of various organs in addition to muscle tissue. These changes can lead to severe impairments and adversely impact the quality of life of affected individuals. The diagnosis and treatment of inflammatory myopathies involve the participation of an interdisciplinary team, due to the complexity of the disease and the high variety of possible signs and symptoms. In this chapter we will discuss the epidemiology and characteristics of the main subtypes of inflammatory myopathies, such as polymyositis, dermatomyositis, necrotizing myopathy, overlap myositis, and myositis of inclusion bodies. Next, we will discuss the existence of crosstalk between inflammatory processes in the oral cavity and their consequences on skeletal muscle. As oral inflammation can increase infiltration of macrophages in muscle tissue and this increase is related to the production of proinflammatory cytokines in this tissue, these cytokines can cause muscle weakness. It is important to consider the prevention of chronic inflammatory processes in order to maintain muscle integrity or even prevent the worsening of the clinical condition of patients with inflammatory muscle diseases.

Keywords: myositis, dermatomyositis, polymyositis, inclusion body myositis, inflammation

1. Introduction

Inflammatory myopathies, also called idiopathic inflammatory myopathy or myositis, are rare conditions characterized by the involvement of various organs in addition to muscle tissue. These changes can lead to severe impairments and adversely impact the quality of life of affected individuals [1, 2].

The diagnosis and treatment of inflammatory myopathies involve the participation of an interdisciplinary team, due to the complexity of the disease and the high variety of possible signs and symptoms. The integration of subspecialties, such as rheumatologist, neurologist, dermatologist, pulmonologist, cardiologist, and physiotherapist, among others, is necessary to achieve the ideal treatment plan. Diagnosis of inflammatory myopathies involves several steps and often requires autoantibody testing and histological evaluation of a muscle tissue biopsy in addition to several other tests, including muscle magnetic resonance imaging and electromyography. Typical symptoms of inflammatory myopathies include muscle weakness in the arms and legs, which may manifest in a few days or even several weeks. Muscular weakness is reflected in difficulties in performing daily activities such as walking, climbing stairs, or lifting an object above the head. In addition to muscle weakness, it is observed that pain is also a frequent detectable symptom in a patient with inflammatory myopathies. Laboratory tests usually show a significant increase of creatine kinase and elevation in the concentration of liver enzymes that suggest the occurrence of damage to muscle cells [1, 3].

The adverse impact on quality of life highlights the importance of performing an accurate and reliable diagnosis from the combination of clinical and laboratory findings to establish the appropriate treatment for each individual [1, 2].

In this chapter we will discuss the epidemiology and subtypes of inflammatory myopathies. Next, we will discuss the existence of crosstalk between inflammatory processes in the oral cavity and their consequences on skeletal muscle.

2. Inflammatory myopathies

2.1 Epidemiology of inflammatory muscle diseases

All myositis subtypes can be considered rare diseases due to their relatively low prevalence. Studies indicate that overlap myositis represents the subtype of the disease that affects the largest number of people, comprising about half of the cases registered. Dermatomyositis accounts for more than a third of the cases of the disease and presents a prevalence of approximately 1–6 patients per 100,000 people in the United States [4–6].

It is important to emphasize that obtaining accurate epidemiological data is extremely difficult due to the different diagnostic criteria adopted in each study. Therefore, the information provided by the publications should be examined and evaluated with caution and attention [7].

A large study conducted from the analysis of 3067 patients from Belgium, China, Czech Republic, Hungary, Italy, Mexico, Norway, Sweden, Switzerland, the United Kingdom (UK), and Vietnam who were registered in the Euromyositis Registry demonstrated that the dermatomyositis was the most common disorder with 31% of the cases [7].

Data on the prevalence of necrotizing myopathy suggest that this subtype of the disease accounts for approximately one-fifth of the reported cases of inflammatory muscle diseases [4–6].

The information regarding the epidemiology of polymyositis varies and depends on the methodology and location of the study ranging from the largest fraction with prevalence of approximately 10 cases per 100,000 people in the United States [1–3], 27% in the Euromyositis Registry [7], to the rarest subtype that should be diagnosed only by exclusion [4–6].

Currently there is some consensus that overlap myositis, necrotizing myopathy, and dermatomyositis represent about 90% of the cases of inflammatory muscle diseases [4–6]. It is estimated that the inclusion body myositis occurs with a prevalence of up to 14 per million people [8].

2.2 Dermatomyositis

Dermatomyositis is typically characterized by the development of proximal muscle weakness and cutaneous manifestations that may arise over a period of weeks to months. However, there are cases in which muscular impairment is not significant without signs and symptoms of muscle weakness, elevated muscle enzymes or changes in electromyography, magnetic resonance imaging (MRI), and muscle biopsy [9].

Skin signs frequently seen in dermatomyositis include an exacerbated periorbital rash with edematous features and erythematous lesions involving the extensor surfaces of the joints. In some cases, myalgia and pruritus may also be observed as important symptoms of the disease. Muscle enzyme concentrations tend to be elevated, and electromyography commonly shows a myopathic pattern [10]. Intramuscular T2 hyperintensities resulting from inflammation or muscle necrosis can be observed on MRI. Dermatomyositis may present a characteristic less frequently observed in other types of inflammatory myopathies, which involves the presence of T2 hyperintensities around individual muscles due to fascial involvement [11].

Muscular biopsies in patients with dermatomyositis have perifascicular atrophy as a feature of high specificity [12]. Evidences show that the expression of perifascicular human myxovirus resistance protein 1 and retinoic acid-inducible gene 1 have higher diagnostic sensitivity than perifascicular atrophy with equivalent specificity [13]. Muscular biopsies of dermatomyositis patients usually present cellular infiltrates composed of plasmacytoid dendritic cells, B cells, CD4 T cells, and macrophages. These cells usually involve medium-sized blood vessels and invade the perimysium [14]. However, it is possible that dermatomyositis biopsy does not present this cellular infiltrate. Predominantly, necrotic pathologically indistinguishable from immune-mediated necrotizing myopathy may be observed. Some early features of dermatomyositis involve deposition of membrane attack complex and presence of microtubular inclusions on intramuscular capillaries [11]. In addition, like other inflammatory myopathies, class-1 major histocompatibility complex (MHC) is generally upregulated in the sarcolemma of muscle fibers. In patients with dermatomyositis, class-1 MHC upregulation and other pathological findings may be characteristically prominent in perifascicular regions [14].

Studies have shown that dermatomyositis autoantibody can be found in a considerable proportion of patients with dermatomyositis [15]. Typical features of dermatomyositis, including proximal muscle weakness and prominent cutaneous manifestations have been associated with the presence of autoantibodies recognizing the nuclear antigen Mi2 [16]. Patients with dermatomyositis and autoantibodies that recognize nuclear matrix protein (NXP) 2 are more predisposed to be affected by proximal and distal muscular weakness, subcutaneous edema, and dysphagia [17].

Patients with dermatomyositis who are positive for anti-NXP2 or anti-transcription intermediary factor (TIF)-1 autoantibodies are at increased risk for malignancy development; thus making comprehensive cancer screening 13–15 or positron emission tomography–computed tomography (PET-CT) scans is extremely important in these cases [18]. In cases of dermatomyositis patients who have autoantibodies recognizing the small ubiquitin-like modifier activating enzyme or melanoma differentiation-associated gene 5 (MDA5), it is observed that cutaneous tissue impairment is more prominent than in muscle. In addition to most commonly present cutaneous manifestations, these patients may develop ulcerous lesions on the flexor surface of the fingers and palm [19, 20].

Most patients with anti-MDA5 autoantibodies are hypomyopathic or amyopathic. In addition, it should be noted that unlike patients with other autoantibodies of dermatomyositis, those who are anti-MDA5 positive often develop an aggressive form of interstitial lung disease, reinforcing the importance of assessment through periodic lung function tests and high-resolution computed tomography [20–22].

Although the etiology of dermatomyositis is not fully elucidated, it is suggested that a combination of genetic risk factors and exposure to environmental factors may trigger the disease. In this sense, several immunogenetic risk factors, including certain class-2 human leukocyte antigen (HLA) alleles, have been implicated in dermatomyositis pathogenesis [23]. Studies suggest that exposure to ultraviolet light may also be considered an important risk factor for the development of dermatomyositis [24].

Regardless of the origin of dermatomyositis, it is not known which mechanisms are involved in the development of muscle damage and weakness. Studies suggest that muscle damage may result from hypoperfusion due to endothelial destruction [14]. In addition, the presence of plasmacytoid dendritic cells, along with the increase in expression of type-1 interferon-inducible proteins in the perifascicular area, suggests that interferon may mediate perifascicular atrophy [12, 25].

Overlap myositis is being recognized as an individual form of myositis. This myositis manifests itself without a rash typical of dermatomyositis, with prominent pathologic changes in the perifascicular, interfascicular, and perimysial regions, and is frequently associated with anti-synthetase antibodies [2].

Laboratory evaluation shows a significant elevation of muscle enzymes including creatine kinase (CK), which is generally present [3]. Approximately 30% of patients with myositis were positive for Jo-1O antibody (most common of the eight anti-synthetase antibodies) [26].

2.3 Polymyositis

Polymyositis is a rare disease, which belongs to the various idiopathic inflammatory myopathies. It is estimated that the incidence of polymyositis is 5% of all cases of myositis [2, 5, 27]. Polymyositis consists of muscle weakness, elevated creatine phosphokinase concentrations, and myopathic electromyography features [2]. However, rash or other signs of skin inflammation do not occur in polymyositis. Therefore, its diagnosis is by exclusion [3].

Histopathological hallmarks of polymyositis include invasion of endomysial cytotoxic CD8 T cells and widespread upregulation of class I MHC in muscle fibers [2, 24]. Polymyositis is a chronic, degenerative disease that has no cure. The treatment consists in the relief of the symptoms with the use of corticosteroids, such as prednisone, intravenous glucocorticoids (when weakness at onset is severe or rapidly worsening), azathioprine, methotrexate, mycophenolate, cyclosporine, and intravenous immune globulin [3].

2.4 Inclusion body myositis

Inclusion body myositis is a very common disease among inflammatory myopathies affecting mainly men from the age of 50. The disease begins insidiously and develops over a period of years, sometimes asymmetrically; it may begin with unilateral affection of a leg or arm, progress steadily, and lead to deep muscular atrophy [2]. Laboratory evaluation shows that an elevated CK is much blander. Skin changes are not present [3].

There is a higher mortality rate in patients with inclusion body myositis, since muscle weakness (long flexors of the fingers, quadriceps, anterior tibial, and, to a lesser extent, all other muscles of the arms and legs) usually leads to harmful falls and dysphagia can cause aspiration pneumonia [3].

The antibody, identified a few years ago, that is present in inclusion body myositis is cN1A (5NT1A/5NTC1A) [3]. The frequency of this antibody is about 30%; other forms of myositis such as dermatomyositis and other conditions such as Sjögren's syndrome and systemic lupus erythematosus (SLE) were also positive even in the absence of any muscle symptoms [3, 28, 29]. Study suggested that the presence of cN1A is associated with a more severe course of disease, dysphagia [3, 30], and increased mortality [3, 31]. However, in another study in German patients, the presence of cN1A did not correlate with the severity of dysphagia or muscle impairment [3, 32].

In the histopathological hallmarks, the distribution and the immunophenotypic profile of the inflammatory cells are similar to those seen in polymyositis macrophages and CD8+ T cells which invade nonnecrotic muscle fibers that express MHC class I antigen on the sarcolemma [33], signs of protein accumulation by detection of amyloid (Congo red, thioflavin S, immunohistochemistry for p62 or TDP-43), detection of tubulofilaments on EM, vacuoles and signs of mitochondrial damage as evidenced by histochemical proof of COX-deficient muscle fibers, and paracrystalline inclusions [3, 34, 35].

2.5 Immune-mediated necrotizing myopathy

Immune-mediated necrotizing myopathy is an acute or subacute proximal weakness of the arms and legs, most prominent in the lower limbs [3]. It often affects adults, but it can also occur in children [3]. The progression of the disease is constantly more rapid and severe compared to other myopathies (dermatomyositis and polymyositis) [3]. Laboratory evaluation shows very high muscle enzymes, with an elevated CK of 20–50 times [3]. Neck muscle weakness and dysphagia are common [3].

Approximately 10–20% of patients with immune-mediated necrotizing myopathy have anti-signal recognition particle (SRP); however its detection varies from 0 to 54% [36]. This antibody may be associated with cardiomyopathy and a severe disease with muscle atrophy, interstitial lung disease, and dysphagia [37, 38]. Another antibody that has been identified is reductase (HMGCR) antibody; its detection in certain cohorts was 60% [39].

Histopathological hallmarks in necrotizing myopathy show dispersed necrotic myofibers of varying degrees; moderate and predominantly MHC class I focal regulation, particularly in areas with necrotic fibers; and complement binding to the sarcolemma [2, 3, 40–42]. Some inflammatory T cells and other immune cells may be present around these focal points, but there are no primary inflammatory lesions. Necrotic fibers typically exhibit a secondary invasion by macrophages to clean the cell debris [3].

2.6 Crosstalk between oral cavity and skeletal muscle

In addition to these inflammatory muscular diseases mentioned above, a localized inflammation at a distance from the skeletal muscle may promote change in this tissue. Recent study proposed the existence of crosstalk between oral cavity and skeletal muscle [43]. The researchers induced oral inflammation in rats and observed that the skeletal muscle was affected by increased infiltration of macrophages, which was suggested by the authors as an explanation for the glucose intolerance shown in animals with oral inflammation [43].

Research conducted over the last 15 years has investigated possible mechanisms that cause changes in macrophages polarization and the effects of these changes on insulin signaling in metabolic organs [44]. These cells exhibit a high degree of functional plasticity, so that the nature of an inflammatory trigger, as well as the cytokines present, can determine their polarization and their functional status [44]. In analogy to the nomenclature T-helper cells (Th), Th1 Th2, macrophages can be classified into two distinct phenotypes: type 1 (M1) classically activated and type 2 (M2) alternatively activated [45].

In vitro, these subsets can be induced by stimulation with interferon gamma (IFN-γ) and lipopolysaccharides (LPS) for M1 or interleukin-4 (IL-4) for M2 [46]. The M1/M2 dichotomy is often used to classify macrophages into pro-inflammatory (M1) or anti-inflammatory (M2) [44]. Among the functions performed by the M1 macrophages, tumor necrosis factor-alpha (TNF-α) production is outstanding [47]. Saghizadeh [48] and collaborators observed that diabetic or insulin-resistant patients have increased expression of TNF-α in skeletal muscle when compared to normoglycemic individuals, suggesting that cytokine plays an important role in the pathogenesis of insulin resistance. TNF-α impairs the insulin signal by decreasing the phosphorylation of insulin receptor substrate 1 (IRS-1) in tyrosine residues [49]. In addition, TNF-α can stimulate some serine kinases including IκB kinase (IKK) and c-Jun amino-terminal kinase (JNK), which promote IRS-1 phosphorylation in serine residues, resulting in insulin signal attenuation [50]. On the other hand, M2 macrophages are associated with tissue repair, angiogenesis, reduction of inflammation, and the improvement of insulin signaling in adipose tissue [45, 51]. In addition to the studies that relate obesity to insulin resistance, there are studies in the literature that demonstrate a correlation between this hormonal resistance and inflammatory processes, such as rheumatoid arthritis and oral inflammations [52–54]. In this context, the apical periodontitis (AP), an oral inflammation, stands out. AP occurs as a consequence of various aggressions to the dental pulp, including physical, iatrogenic, infectious, and endodontic traumas. This inflammatory picture can cause a wide variety of immunological responses, in order to protect the dental pulp and periapical regions. The regulation of periapical inflammation is extremely complex, as it involves host mediators, including immunological components such as antibodies, cytokines, arachidonic acid metabolites, and neuropeptides [55]. The characteristic inflammatory process of AP presents different types of gram-negative anaerobic bacteria [56] with LPS in the cell wall [57]. Studies have reported that bacteria which are present in the oral cavity can release LPS into the systemic circulation [58]. This substance has the ability to activate toll-like receptors (TLRs), a cell surface receptor that activates innate immunity and induces inflammatory responses. LPS is a specific ligand for TLR2 and TLR4 but has a higher specificity for TLR4 [59, 60]. When released by gram-negative bacteria, LPS binds to a soluble plasma protein called LPS binding protein. LPS or LPS binding protein [61, 62] binds to the CD14 co-receptor via lipopolysaccharide binding protein (LPB), forming the LPS-CD14 complex. This complex, in turn, is recognized by the TLR4-MD-2 complex, present on the cell surface, which is capable of promoting intracellular recruitment of adapter molecules with N-terminal TIR domain, such as myeloid differentiation primary response 88 (MYD88). This molecule can activate the serine kinases JNK and IKKα/β, which promote activation of the activating proteins-1 (AP-1) and factor nuclear kappa B (NF-κB) transcription factors, respectively [63, 64]. NF-κB regulates the expression of several genes involved in different cellular processes such as inflammatory and immune responses and cell growth and development. In the absence of an NF-κB-activating stimulus, this protein is present in the cytoplasm inactive with an inhibitory protein, IκB [65]. Activation of NF-κB can occur not only by exposure of the cells to LPS but also by the action of inflammatory cytokines (TNF-α and IL-1), activation of T and B lymphocytes, UV radiation, and expression of products [66]. After stimulation, the IKK is phosphorylated and activated. The IKK complex consists of two catalytic subunits, IKK-α and IKK-β, in addition to the NF-kappa-B essential modulator (NEMO) or IKK-γ [67]. After activation, IKK recruits and phosphorylates the IκB that is recognized by the ubiquitin ligase machinery, which leads to its polyubiquitination and consequent degradation. In this way, the NF-κB dimers translocate to the nucleus, binding at specific sites of the deoxyribonucleic acid (DNA) and promoting the transcription of a large number of genes [65, 67].

In addition to activating the IKKα/β/NF-κB pathway, TLRs are capable of activating the JNK pathway [68]. The serine/threonine kinase group called JNK (JNK-1, JNK-2, and JNK-3) belongs to the family of mitogen-activated protein kinase (MAPKs), responsible for the regulation of various cellular functions. This regulation occurs largely because of its ability to control the transcription of specific genes by AP-1 [69]. AP-1 is a transcription factor that, when activated, promotes the expression of genes related to innate immunity [70]. In addition to LPS, the signaling pathway of TLRs can be activated by heat shock proteins [71]. Heat shock proteins

(HSP) are proteins characterized as chaperones because they have an important function in adaptation to stress and cellular protection, acting mainly in the synthesis and protein degradation, besides regulating fundamental cellular processes [72]. The family of HSPs is divided into subfamilies, classified according to the molecular mass, being small HSPs (8–27 kDa) and large HSPs (100–110 kDa), among which stand out HSP90, HSP70, and HSP60 [73]. In addition to its essential functions as a chaperone [74], HSP70 has an anti-inflammatory effect by inhibiting the activation of NF-KB when present in the intracellular environment [75]. However, stimuli such as cell necrosis and bacterial products such as LPS can cause the passage of HSP70 through the membrane into the extracellular environment [76, 77]. Studies have suggested that elevated serum levels of HSP70 may be correlated with cardiovascular disorder, pulmonary fibrosis, renal damage, oxidative stress, and inflammation [78]. The development of these conditions may occur due to the ability of HSP70 to bind to TLR2 and TLR4, promoting the activation of the NF-κB pathway which, as mentioned above, induces the expression of inflammatory mediators related to insulin resistance [79]. Studies suggest that insulin sensitivity may undergo regulatory action by the adaptive immune system [80, 81]. This system is composed of different types of cells, among which the B and T lymphocytes [82] stand out. T lymphocytes are classified into two main classes: helper T lymphocytes, also known as T helper (Th), and cytotoxic T lymphocytes. The "naïve" Th1 lymphocytes, when interacting with antigen presenting cells, undergo activation and can differentiate into different subtypes [83]. The Th1 subtype expresses proinflammatory cytokines, such as TNF-α and IFN-γ; Th2 expresses mainly anti-inflammatory cytokines, such as interleukin-4 (IL-4) and interleukin-13 (IL-13), and regulatory T cells secrete predominantly anti-inflammatory cytokine and transforming growth factor-β (TGF-β) [84]. Th1 cells play a central role in the recruitment of macrophages and induction of insulin resistance in obesity-induced diabetes models. These effects are counterbalanced by the function of Th2 and Treg cells that maintain an anti-inflammatory state and increase insulin sensitivity [85]. Appropriate regulation of Th cells is of extreme importance for the control and prevention of various diseases [86]. An increase or decrease in the Th1 or Th2 subtypes, as well as the cytokines produced by these cells, indicates an imbalance that may be one of the factors responsible for the development of insulin resistance [87]. It is known that insulin resistance is one of the main characteristics of diabetes mellitus [88]. This disease is also closely related to muscle weakness due to altered insulin action [89], standing out that insulin is an important anabolic hormone for protein metabolism [90].

The study performed by Boon et al. [91] with healthy lean individuals observed that only 5 days of hyperlipidic diet promoted increased messenger ribonucleic acid (mRNA) expression of macrophage markers in skeletal muscle and reduced expression of the glucose transporter type 4 (GLUT-4) glucose transporter protein in this tissue. Similarly, Patsouris et al. [92] demonstrated increased macrophage content in skeletal muscle in diabetic patients independently of body mass index (BMI). An increased macrophage content (assessed by F4/80 protein detection) was observed in muscle tissue of rats with AP in the absence of obesity, highlighting the key role of these cells in the etiology of insulin resistance. It should be noted that only F4/80 detection is not able to provide details on M1-type and M2-type macrophage polarization although evidence demonstrates that under obesity conditions, macrophages infiltrated into muscle tissue exhibit phenotype characteristic of M1 polarization [92–95]. The reprogramming of the M1 polarization toward the M2 polarization may represent a promising strategy for the treatment of glycemic homeostasis in patients with diabetes and insulin resistance [44].

As previously reported, inflammation causes insulin resistance. According to Pereira et al., rats with AP had increased IKKα/β and JNK phosphorylation status in gastrocnemius muscle. These results are in agreement with the study of Yaspelkis et al. [96], who observed a higher IKKα/β phosphorylation status in the skeletal muscle of rats treated with a hyperlipidic diet for 12 weeks, and also the study by Todd et al. [97] that identified an increase in JNK activity in the skeletal muscle of rats subjected to 3 weeks of hyperlipidic diet. Kaneto et al. [98] reported that treatment of diabetic rats with JNK inhibitors improved the insulin sensitivity of the animals. Similarly, studies by Yuan et al. [99] and Hundal et al. [100] have reported that inhibition of IKK-β by the administration of salicylates improves insulin action in obese and diabetic human and rats. Furthermore, it has been demonstrated that genetically modified mice, which do not express IKK-β or JNK, are protected from obesity-induced insulin resistance [99, 101–103].

In addition to stimulating inhibitory effects on insulin signal transduction, TNF-α may interact with tumor necrosis factor receptor 1 (TNFR1) in skeletal muscle [104] and thereby stimulate the NF-κB and/or MAPK pathway [105, 106], which are related to the phosphorylation of IKK and JNK and, in their turn, may impair insulin action. Pereira et al. [43] evaluated the plasma concentrations of LPS and HSP70 in AP models. Rats with AP showed a significant increase in both LPS and HSP70 when compared to the control group. Research on diabetes suggests that chronic elevation of LPS levels may play a key role in the development of insulin resistance [107, 108].

Among the possible mechanisms involved in this alteration, we highlight the ability of LPS to bind to the TLR4 receptor, which may trigger the activation of inflammatory signaling pathways related to inhibition of the insulin signal [108]. Another mediator that plays an active role in the modulation of inflammation is the heat shock proteins. The study by Goodman et al. [109] reported higher expression of 44 HSP genes in periapical granulomas compared to healthy periodontal tissues. Elevation of HSP70 plasma concentrations observed in rats with AP may indicate that increased local HSP expression is associated with higher concentrations of this protein in serum [43]. Interestingly, studies have shown that serum concentrations of HSP70 are higher in diabetic patients [110, 111]. Asea et al. [79] reported that HSP70 can bind to the TLR4 receptor, suggesting a possible involvement of this protein in the development of insulin resistance. With regard to the adaptive immunity markers, animals from the AP groups showed an increase in the Th1 response represented by increased T-bet expression in the spleen and elevated plasma concentrations of INF-γ [43]. A study carried out with knockout animals for the T-bet gene treated with hypercaloric diet showed that even with weight gain and increased adiposity, the animals were protected from insulin resistance [112]. The authors attributed the lack of insulin resistance to reduced production of INF-γ. These results are consistent with studies that reported that IFN-γ deficiency may improve glycemic homeostasis under obesity conditions [113–115]. In addition, treatment of adipocytes (3 T3-L1) with interferon gamma (INF-γ) reduces insulin signal and glucose uptake [116]. The functions of Th1 cells are antagonized by the Th2 subpopulation presenting the transcription factor GATA3 and IL-4 as specific markers. The AP in rats promotes a reduction of IL-4 [43]. Chang et al. [117] reported that IL-4 treatment promotes improved insulin sensitivity and glucose tolerance and simultaneously reduces body weight in obese rats. These findings suggest that IL-4 plays beneficial effects on glycemic homeostasis. The role of Th2 cells in insulin sensitivity was demonstrated in the study by Gonzales et al. [118]. In this study, a model of inactivation of Th2 response was developed through the inhibition of the activator of transcription 6 (STAT6) protein, in which it was observed that animals with Th2 response deficiency were more prone to insulin resistance. Thus, the reduction of the Th2 response observed in rats with AP may contribute to the understanding of the mechanisms involved in insulin resistance observed in animals with AP [43, 54].

Studies suggest that TNF-α contributes to age-related muscle loss and that resistance exercise may attenuate this process by suppressing TNF-α expression in skeletal muscle [119]. Other findings demonstrate that decreased muscle strength in diabetic individuals is associated with elevated plasma concentrations of TNF-α and interleukin-6 (IL-6) [120]. Therefore, considering that oral inflammation, such as AP, may increase infiltration of macrophages in muscle tissue and this increase is related to the production of proinflammatory cytokines, it is possible to suggest that prevention of chronic inflammatory oral diseases contributes to the maintenance of muscle integrity.

3. Conclusions

The main subtypes of inflammatory muscular diseases are polymyositis, dermatomyositis, necrotizing myopathy, overlap myositis, and myositis of inclusion bodies. The origin of these diseases is idiopathic, making it difficult to prevent them. As oral inflammation can increase infiltration of macrophages in muscle tissue and this increase is related to the production of proinflammatory cytokines in this tissue, these cytokines can cause muscle weakness. It is important to consider the prevention of chronic inflammatory processes in order to maintain muscle integrity or even prevent the worsening of the clinical condition of patients with inflammatory muscle diseases.

Conflict of interest

The authors declare that there are no conflicts of interest.

Funding

This study was supported by the São Paulo Research Foundation (FAPESP) [grant #2016/24829-2] São Paulo, SP, Brazil.

Appendices and nomenclature

AP	apical periodontitis
AP-1	activating proteins-1
BMI	body mass index
CD14	cluster of differentiation 14
CD4	cluster of differentiation 4
CD8	cluster of differentiation 8
CK	creatine kinase
COX	cyclooxygenase
DNA	deoxyribonucleic acid
GLUT4	glucose transporter type 4
HLA	human leukocyte antigen
HMGCR	3-hydroxy-3-methylglutaryl-CoA reductase
HSP	heat shock proteins
IFN-γ	interferon gamma
IKK	IkB kinase
IL-13	interleukin-13
IL-4	interleukin-4
IL-6	interleukin-6

IR	insulin resistance
IRS-1	insulin receptor substrate 1
JNK	c-jun amino-terminal kinase
LPB	lipopolysaccharide binding protein
LPS	lipopolysaccharides
M1	M1-type macrophage polarization
M2	M2-type macrophage polarization
MAPKs	mitogen-activated protein kinase
MDA5	melanoma differentiation-associated gene 5
MHC	major histocompatibility complex
MRI	magnetic resonance imaging
mRNA	messenger ribonucleic acid
MyD88	myeloid differentiation primary response 88
NEMO	NF-kappa-B essential modulator
NF-κB	factor nuclear kappa B
NXP	nuclear matrix protein
PET-CT	positron emission tomography–computed tomography
SLE	systemic lupus erythematosus
SRP	signal recognition particle
STAT6	activator of transcription 6
TDP-43	transactive DNA-binding protein 43
TGF-β	transforming growth factor-β
Th	T-helper
TIF	transcription intermediary factor
TLR	toll-like receptors
TNFR1	tumor necrosis factor receptor 1
TNF-α	tumor necrosis factor-alpha
UK	United Kingdom
USA	United States of America

Author details

Doris Hissako Sumida[1]*, Fernando Yamamoto Chiba[2]
and Maria Sara de Lima Coutinho Mattera[1]

1 Multicenter Post-Graduate Program in Physiological Sciences (SBFis), Department of Basic Sciences, School of Dentistry, São Paulo State University (UNESP), Araçatuba, Brazil

2 Department of Child and Social Dentistry, School of Dentistry, São Paulo State University (UNESP), Araçatuba, Brazil

*Address all correspondence to: doris.hissako@unesp.br

References

[1] Carstens PO, Schmidt J. Diagnosis, pathogenesis and treatment of myositis: Recent advances. Clinical and Experimental Immunology. 2014;**175**:349-358. DOI: 10.1111/cei.12194

[2] Dalakas MC. Inflammatory muscle diseases. The New England Journal of Medicine. 2015;**372**:1734-1747. DOI: 10.1056/NEJMra1402225

[3] Schmidt J. Current classification and management of inflammatory myopathies. Journal of Neuromuscular Diseases. 2018;**5**:109-129. DOI: 10.3233/JND-180308

[4] Furst DE, Amato AA, Iorga Ş, Gajria K, Fernandes AW. Epidemiology of adult idiopathic inflammatory myopathies in a U.S. managed care plan. Muscle & Nerve. 2012;**45**:676-683. DOI: 10.1002/mus.23302

[5] Senécal JL, Raynauld JP, Troyanov Y. Editorial: A new classification of adult autoimmune myositis. Arthritis & Rhematology. 2017;**69**:878-884. DOI: 10.1002/art.40063

[6] Van der Meulen MF, Bronner IM, Hoogendijk JE, Burger H, van Venrooij WJ, Voskuyl AE, et al. Polymyositis: An overdiagnosed entity. Neurology. 2003;**61**:316-321. DOI: 10.1212/WNL.61.3.316

[7] Lilleker JB, Vencovsky J, Wang G, Wedderburn LR, Diederichsen LP, Schmidt J, et al. The EuroMyositis registry: An international collaborative tool to facilitate myositis research. Annals of the Rheumatic Diseases. 2018;**77**:30-39. DOI: 10.1136/annrheumdis-2017-211868

[8] Needham M, Corbett A, Day T, Christiansen F, Fabian V, Mastaglia FL. Prevalence of sporadic inclusion body myositis and factors contributing to delayed diagnosis. Journal of Clinical Neuroscience. 2008;**15**:1350-1353. DOI: 10.1016/j.jocn.2008.01.011

[9] Selva-O'Callaghan A, Pinal-Fernandez I, Trallero-Araguás E, Milisenda JC, Grau-Junyent JM, Mammen AL. Classification and management of adult inflammatory myopathies. Lancet Neurology. 2018;**17**:816-828. DOI: 10.1016/S1474-4422(18)30254-0

[10] Paganoni S, Amato A. Electrodiagnostic evaluation of myopathies. Physical Medicine and Rehabilitation Clinics of North America. 2013;**24**:193-207. DOI: 10.1016/j.pmr.2012.08.017

[11] Pinal-Fernandez I, Casal-Dominguez M, Carrino JA, Lahouti AH, Basharat P, Albayda J, et al. Thigh muscle MRI in immune-mediated necrotising myopathy: Extensive oedema, early muscle damage and role of anti-SRP autoantibodies as a marker of severity. Annals of the Rheumatic Diseases. 2017;**76**:1-7. DOI: 10.1136/annrheumdis-2016-210198

[12] Suárez-Calvet X, Gallardo E, Pinal-Fernandez I, De Luna N, Lleixà C, Díaz-Manera J, et al. RIG-I expression in perifascicular myofibers is a reliable biomarker of dermatomyositis. Arthritis Research & Therapy. 2017;**19**:174. DOI: 10.1186/s13075-017-1383-0

[13] Uruha A, Nishikawa A, Tsuburaya RS, Hamanaka K, Kuwana M, Watanabe Y, et al. Sarcoplasmic MxA expression: A valuable marker of dermatomyositis. Neurology. 2017;**88**:493-500. DOI: 10.1212/WNL.0000000000003568

[14] Dalakas MC. Muscle biopsy findings in inflammatory myopathies. Rheumatic Diseases Clinics of North America. 2002;**28**:779-798. DOI: 10.1016/S0889-857X (02)00030-3

[15] Betteridge Z, McHugh N. Myositis-specific autoantibodies: An important tool to support diagnosis of myositis. Journal of Internal Medicine. 2016;**280**:8-23. DOI: 10.1111/joim.12451

[16] Ghirardello A, Zampieri S, Iaccarino L, Tarricone E, Bendo R, Gambari PF, et al. Anti-Mi-2 antibodies. Autoimmunity. 2005;**38**:79-83. DOI: 10.1080/08916930400022681

[17] Albayda J, Pinal-Fernandez I, Huang W, Parks C, Paik J, Casciola-Rosen L, et al. Antinuclear matrix protein 2 autoantibodies and edema, muscle disease, and malignancy risk in dermatomyositis patients. Arthritis Care & Research. 2017;**69**:1771-1776. DOI: 10.1002/acr.23188

[18] Selva-O'Callaghan A, Grau JM, Gamez-Cenzano C, Vidaller-Palacin A, Martinez-Gomez X, Trallero-Araguas E, et al. Conventional cancer screening versus PET/CT in dermatomyositis/polymyositis. The American Journal of Medicine. 2010;**123**:558-562. DOI: 10.1016/j.amjmed.2009.11.012

[19] Ge Y, Lu X, Shu X, Peng Q, Wang G. Clinical characteristics of anti-SAE antibodies in Chinese patients with dermatomyositis in comparison with different patient cohorts. Scientific Reports. 2017;**7**:188. DOI: 10.1038/s41598-017-00240-6

[20] Narang NS, Casciola-Rosen L, Li S, Chung L, Fiorentino DF. Cutaneous ulceration in dermatomyositis: Association with anti-melanoma differentiation-associated gene 5 antibodies and interstitial lung disease. Arthritis Care & Research. 2015;**67**:667-672. DOI: 10.1002/acr.22498

[21] Labrador-Horrillo M, Martinez MA, Selva-O'Callaghan A, Trallero-Araguas E, Balada E, Vilardell-Tarres M, et al. Anti-MDA5 antibodies in a large Mediterranean population of adults with dermatomyositis. Journal of Immunology Research. 2014;**2014**:290-797. DOI: 10.1155/2014/290797

[22] Sato S, Hoshino K, Satoh T, Fujita T, Kawakami Y, Kuwana M. RNA helicase encoded by melanoma differentiation-associated gene 5 is a major autoantigen in patients with clinically amyopathic dermatomyositis: Association with rapidly progressive interstitial lung disease. Arthritis and Rheumatism. 2009;**60**:2193-2200. DOI: 10.1002/art.24621

[23] Miller FW, Chen W, O'Hanlon TP, Cooper RG, Vencovsky J, Rider LG, et al. Genome-wide association study identifies HLA 8.1 ancestral haplotype alleles as major genetic risk factors for myositis phenotypes. Genes and Immunity. 2015;**16**:470-480. DOI: 10.1038/gene.2015.28

[24] Mamyrova G, Rider LG, Ehrlich A, Jones O, Pachman LM, Nickeson R, et al. Environmental factors associated with disease flare in juvenile and adult dermatomyositis. Rheumatology. 2017;**56**:1342-1347. DOI: 10.1093/rheumatology/kex162

[25] Greenberg SA, Pinkus JL, Pinkus GS, Burleson T, Sanoudou D, Tawil R, et al. Interferon-alpha/beta-mediated innate immune mechanisms in dermatomyositis. Annals of Neurology. 2005;**57**:664-678. DOI: 10.1002/ana.20464

[26] Lega JC, Fabien N, Reynaud Q, Durieu I, Durupt S, Dutertre M, et al. The clinical phenotype associated with myositis-specific and associated autoantibodies: A meta-analysis revisiting the so-called antisynthetase syndrome. Autoimmunity Reviews. 2014;**13**:883-891. DOI: 10.1016/j.autrev.2014.03.004

[27] Benveniste O, Stenzel W, Allenbach Y. Advances in serological diagnostics of inflammatory myopathies. Current Opinion in Neurology.

2016;**29**:662-673. DOI: 10.1097/WCO.0000000000000376

[28] Lloyd TE, Christopher-Stine L, Pinal-Fernandez I, Tiniakou E, Petri M, Baer A, et al. Cytosolic 5'-nucleotidase 1A As a target of circulating autoantibodies in autoimmune diseases. Arthritis Care & Research. 2016;**68**:66-71. DOI: 10.1002/acr.22600

[29] Herbert MK, Stammen-Vogelzangs J, Verbeek MM, Rietveld A, Lundberg IE, Chinoy H, et al. Disease specificity of autoantibodies to cytosolic 5′-nucleotidase 1A in sporadic inclusion body myositis versus known autoimmune diseases. Annals of the Rheumatic Diseases. 2016;**75**:696-701. DOI: 10.1136/annrheumdis-2014-206691

[30] Goyal NA, Cash TM, Alam U, Enam S, Tierney P, Araujo N, et al. Seropositivity for NT5c1A antibody in sporadic inclusion body myositis predicts more severe motor, bulbar and respiratory involvement. Journal of Neurology, Neurosurgery, and Psychiatry. 2016;**87**:373-378. DOI: 10.1136/jnnp-2014-310008

[31] Lilleker JB, Rietveld A, Pye SR, Mariampillai K, Benveniste O, Peeters MT, et al. Cytosolic 5′-nucleotidase 1A autoantibody profile and clinical characteristics in inclusion body myositis. Annals of the Rheumatic Diseases. 2017;**76**:862-868. DOI: 10.1136/annrheumdis-2016-210282

[32] Olthoff A, Carstens PO, Zhang S, von FE, Friede T, Lotz J, et al. Evaluation of dysphagia by novel real-time MRI. Neurology. 2016;**87**:2132-2138. DOI: 10.1212/WNL.0000000000003337

[33] Vattemi G, Mirabella M, Guglielmi V, Lucchini M, Tomelleri G, Ghirardello A, et al. Muscle biopsy features of idiopathic inflammatory myopathies and differential diagnosis. Autoimmunity Highlights. 2014;**5**:77-85. DOI: 10.1007/s13317-014-0062-2

[34] Ikenaga C, Kubota A, Kadoya M, Taira K, Uchio N, Hida A, et al. Clinicopathologic features of myositis patients with CD8-MHC-1 complex pathology. Neurology. 2017;**89**:1060-1068. DOI: 10.1212/WNL.0000000000004333

[35] Chahin N, Engel AG. Correlation of muscle biopsy, clinical course, and outcome in PM and sporadic IBM. Neurology. 2008;**70**:418-424. DOI: 10.1212/01.wnl.0000277527.69388.fe

[36] Satoh M, Tanaka S, Ceribelli A, Calise SJ, Chan EKA. Comprehensive overview on myositis-specific antibodies: New and old biomarkers in idiopathic inflammatory myopathy. Clinical Reviews in Allergy and Immunology. 2017;**52**:1-19. DOI: 10.1007/s12016-015-8510-y

[37] Targoff IN, Johnson AE, Miller FW. Antibody to signal recognition particle in polymyositis. Arthritis and Rheumatism. 1990;**33**:1361-1370. DOI: 10.1002/art.1780330908

[38] Suzuki S, Nishikawa A, Kuwana M, Nishimura H, Watanabe Y, Nakahara J, et al. Inflammatory myopathy with anti-signal recognition particle antibodies: Case series of 100 patients. Orphanet Journal of Rare Diseases. 2015;**10**:61. DOI: 10.1186/s13023-015-0277-y

[39] Christopher-Stine L, Casciola-Rosen LA, Hong G, Chung T, Corse AM, Mammen AL. A novel autoantibody recognizing 200-kd and 100-kd proteins is associated with an immune-mediated necrotizing myopathy. Arthritis and Rheumatism. 2010;**62**:2757-2766. DOI: 10.1002/art.27572

[40] Hoogendijk JE, Amato AA, Lecky BR, Choy EH, Lundberg IE, Rose MR, et al. 119th ENMC international workshop: Trial design in adult idiopathic inflammatory myopathies,

with the exception of inclusion body myositis, 10-12 October 2003, Naarden, The Netherlands. Neuromuscular Disorders. 2004;**14**:337-345. DOI: 10.1016/j.nmd.2004.02.006

[41] Allenbach Y, Benveniste O, Goebel HH, Stenzel W. Integrated classification of inflammatory myopathies. Neuropathology and Applied Neurobiology. 2017;**43**:62-81. DOI: 10.1111/nan.12380

[42] Stenzel W, Goebel HH, Aronica E. Review: Immune-mediated necrotizing myopathies–a heterogeneous group of diseases with specific myopathological features. Neuropathology and Applied Neurobiology. 2012;**38**:632-646. DOI: 10.1111/j.1365-2990.2012.01302.x

[43] Pereira RF, Cintra LTA, Tessarin GWL, Chiba FY, de Lima Coutinho Mattera MS, Scaramele NF, et al. Periapical lesions increase macrophage infiltration and inflammatory signaling in muscle tissue of rats. Journal of Endodontia. 2017;**43**:982-988. DOI: 10.1016/j.joen.2017.01.030

[44] Lauterbach MA, Wunderlich FT. Macrophage function in obesity-induced inflammation and insulin resistance. Pflügers Archiv. 2017;**469**:385-396. DOI: 10.1007/s00424-017-1955-5

[45] Lumeng CN, Bodzin JL, Saltiel AR. Obesity induces a phenotypic switch in adipose tissue macrophage polarization. The Journal of Clinical Investigation. 2007;**117**:175-184. DOI: 10.1172/JCI29881

[46] Ginhoux F, Schultze JL, Murray PJ, Ochando J, Biswas SK. New insights into the multidimensional concept of macrophage ontogeny, activation and function. Nature Immunology. 2016;**17**:39-40. DOI: 10.1038/ni.3324

[47] Metzger Z. Macrophages in periapical lesions. Endodontics & Dental Traumatology. 2000;**16**:1-8

[48] Saghizadeh M, Ong JM, Garvey WT, Henry RR, Kern PA. The expression of TNF a by human muscle. The Journal of Clinical Investigation. 1996;**97**: 1111-1116. DOI: 10.1172/JCI118504

[49] Hotamisligil GS. IRS-1-mediated inhibition of insulin receptor tyrosine kinase activity in TNF-alpha- and obesity-induced insulin resistance. Science. 1996;**271**:665-658. DOI: 10.1126/science.271.5249.665

[50] Osborn O, Olefsky JM. The cellular and signaling networks linking the immune system and metabolism in disease. Nature Medicine. 2012;**18**:363-374. DOI: 10.1038/nm.2627

[51] Nguyen KD, Qiu Y, Cui X, Goh YP, Mwangi J, David T, et al. Alternatively activated macrophages produce catecholamines to sustain adaptive thermogenesis. Nature. 2011;**20**:104-108. DOI: 10.1038/nature10653

[52] Silber SJ. Apparent fertility of human spermatozoa from the caput epididymitis. Journal of Andrology. 1989;**4**:263-269. DOI: 10.1002/j.1939-4640.1989.tb00098.x

[53] Giles JT, Danielides S, Szklo M, Post WS, Blumenthal RS, Petri M, et al. Insulin resistance in rheumatoid arthritis: Disease-related indicators and associations with the presence and progression of subclinical atherosclerosis. Arthritis & Rhematology. 2015;**67**:626-636. DOI: 10.1002/art.38986

[54] Astolphi RD, Curbete MM, Colombo NH, Shirakashi DJ, Chiba FY, Prieto AK, et al. Periapical lesions decrease insulin signal and cause insulin resistance. Journal of Endodontia. 2013;**39**:648-652. DOI: 10.1016/j.joen.2012.12.031

[55] Stashenko P, Teles R, D'Souza R. Periapical inflammatory responses and their modulation. Critical

Reviews in Oral Biology and Medicine. 1998;**9**:498-521. DOI: 10.1177/10454411980090040701

[56] Sundqvist G. Ecology of the root canal flora. Journal of Endodontia. 1992;**18**:427-430. DOI: 10.1016/S0099-2399(06)80842-3

[57] Wolff SM. Biological effects of bacterial endotoxins in man. The Journal of Infectious Diseases. 1973;**128**:259-264. DOI: 10.1093/infdis/128.Supplement_1.S259

[58] Geerts SO, Nys M, De MP, Charpentier J, Albert A, Legrand V, et al. Systemic release of endotoxins induced by gentle mastication: Association with periodontitis severity. Journal of Periodontology. 2002;**73**:73-78. DOI: 10.1902/jop.2002.73.1.73

[59] Medzhitov R. Toll-like receptors and innate immunity. Nature Reviews Immunology. 2001;**1**:135-145. DOI: 10.1038/35100529

[60] Takeuchi O, Sato S, Horiuchi T, Hoshino K, Takeda K, Dong Z, et al. Cutting edge: Role of toll-like receptor 1 in mediating immune response to microbial lipoproteins. Journal of Immunology. 2002;**169**:10-14. DOI: 10.4049/jimmunol.169.1.10

[61] Poltorak A, He X, Smirnova I, Liu MY, Van Huffel C, Du X, et al. Defective LPS signaling in C2H/HeJ and C57BL/10ScCr mice: Mutations in Tlr4 gene. Science. 1998;**282**:2085-2088. DOI: 10.1126/science.282.5396.2085

[62] Akira S, Uematsu S, Takeuchi O. Pathogen recognition and innate immunity. Cell. 2006;**124**:783-801. DOI: 10.1016/j.cell.2006.02.015

[63] Ohnishi T, Muroi M, Tanamoto K. The lipopolysaccharide-recognition mechanism in cells expressing TLR4 and CD14 but lacking MD-2. FEMS Immunology and Medical Microbiology. 2007;**51**:84-91. DOI: 10.1111/j.1574-695X.2007.00281.x

[64] Pålsson-McDermott EM, O'Neill LA. Signal transduction by the lipopolysaccharide receptor, toll-like receptor-4. Immunology. 2004;**113**:153-162. DOI: 10.1111/j.1365-2567.2004.01976.x

[65] Napetschnig J, Wu H. Molecular basis of NF-κB signaling. Annual Review of Biophysics. 2013;**42**:443-468. DOI: 10.1146/annurev-biophys-083012-130338

[66] Baldwin AS Jr. The NF-kappa B and I kappa B proteins: New discoveries and insights. Annual Review of Immunology. 1996;**14**:649-683. DOI: 10.1146/annurev.immunol.14.1.649

[67] Li Q, Verma IM. NF-kb regulation in the immune system. Nature Reviews. Immunology. 2002;**12**:975. DOI: 10.1038/nri910

[68] Shoelson SE, Lee J, Goldfine AB. Inflammation and insulin resistance. The Journal of Clinical Investigation. 2006;**116**:1793-1801. DOI: 10.1172/JCI29069

[69] Wellen KE, Hotamisligil GS. Inflammation, stress, and diabetes. The Journal of Clinical Investigation. 2005;**115**:1111-1119. DOI: 10.1172/JCI25102

[70] Bennett BL, Satoh Y, Lewis AJ. JNK: A new therapeutic target for diabetes. Current Opinion in Pharmacology. 2003;**3**:420-425. DOI: 10.1016/S1471-4892(03)00068-7

[71] Fang H, Wu Y, Huang X, Wang W, Ang B, Cao X, et al. Toll-like receptor 4 (TLR4) is essential for Hsp70-like protein 1 (HSP70L1) to activate dendritic cells and induce Th1 response. The Journal of Biological Chemistry. 2011;**286**:30393-30400. DOI: 10.1074/jbc.M111.266528

[72] Lianos GD, Alexiou GA, Mangano A, Mangano A, Rausei S, Boni L, et al. The role of heat shock proteins in câncer. Cancer Letters. 2015;**360**:114-118. DOI: 10.1016/j.canlet.2015.02.026

[73] Lindquist S, Craig EA. The heat shock proteins. Annual Review of Genetics. 1988;**22**:631-637

[74] Pilon M, Schekman R. Protein translocation: How Hsp70 pulls it off. Cell. 1999;**97**:679-682. DOI: 10.1016/S0092-8674(00)80780-1

[75] Heck TG, Schöler CM, de Bittencourt PI. HSP70 expression: Does it a novel fatigue signalling factor from immune system to the brain? Cell Biochemistry and Function. 2011;**29**:215-226. DOI: 10.1002/cbf.1739

[76] Davies EL, Bacelar MM, Marshall MJ, Johnson E, Wardle TD, Andrew SM, et al. Heat shock proteins form part of a danger signal cascade in response to lipopolysaccharide and GroEL. Clinical and Experimental Immunology. 2006;**145**:183-189. DOI: 10.1111/j.1365-2249.2006.03109.x

[77] Mambula SS, Stevenson MA, Ogawa K, Calderwood SK. Mechanisms for Hsp70 secretion: Crossing membranes without a leader. Methods. 2007;**43**:168-175. DOI: 10.1016/j.ymeth.2007.06.009

[78] Ogawa F, Shimizu K, Hara T, Muroi E, Hasegawa M, Takehara K, et al. Serum levels of heat shock protein 70, a biomarker of cellular stress, are elevated in patients with systemic sclerosis: Association with fibrosis and vascular damage. Clinical and Experimental Rheumatology. 2008;**26**:659-662

[79] Asea A, Rehli M, Kabingu E, Boch JA, Bare O, Auron PE, et al. Novel signal transduction pathway utilized by extracellular HSP70: Role of toll-like receptor (TLR)2 and TLR4. The Journal of Biological Chemistry. 26 Apr 2002;**277**(17):15028-15034. DOI: 10.1074/jbc.M200497200

[80] Bronsart LL, Contag CH. A role of the adaptive immune system in glucose homeostasis. BMJ Open Diabetes Research & Care. 2016;**4**:e000136. DOI: 10.1136/bmjdrc-2015-000136

[81] Winer S, Winer DA. The adaptive immune system as a fundamental regulator of adipose tissue inflammation and insulin resistance. Immunology and Cell Biology. 2012;**90**:755-762. DOI: 10.1038/icb.2011.110

[82] Corthay A. How do regulatory T cells work. Scandinavian Journal of Immunology. 2009;**70**:326-336. DOI: 10.1111/j.1365-3083.2009.02308.x

[83] Antignano F, Zaph C. Regulation of CD4 T-cell differentiation and inflammation by repressive histone methylation. Immunology and Cell Biology. 2015;**93**:245-252. DOI: 10.1038/icb.2014.115

[84] Hansson GK. Atherosclerosis-an immune disease: The anitschkov lecture 2007. Atherosclerosis. 2009;**202**:2-10. DOI: 10.1016/j.atherosclerosis.2008.08.039

[85] Wali JA, Thomas HE, Sutherland AP. Linking obesity with type 2 diabetes: The role of T-bet. Diabetes, Metabolic Syndrome and Obesity: Targets and Therapy. 2014;7:331-340. DOI: 10.2147/DMSO.S51432

[86] Zhao G, Zhou S, Davie A, Su Q. Effects of moderate and high intensity exercise on T1/T2 balance. Exercise Immunology Review. 2012;**18**:98-114

[87] Deiuliis J, Shah Z, Shah N, Needleman B, Mikami D, Narula V, et al. Visceral adipose inflammation in obesity is associated with critical alterations in T regulatory cell numbers. PLoS One. 2011;**6**:16376. DOI: 10.1371/journal.pone.0016376

[88] Taylor R. Insulin resistance and type 2 diabetes. Diabetes. 2012;**61**:778-779. DOI: 10.2337/db12-0073

[89] Perry BD, Caldow MK, Brennan-Speranzam TC, Sbaraglia M, Jerums G, Garnham A, et al. Muscle atrophy in patients with type 2 diabetes mellitus: Roles of inflammatory pathways, physical activity and exercise. Exercise Immunology Review. 2016;**22**:94-108

[90] Qaid MM, Abdelrahman MM. Role of insulin and other related hormones in energy metabolism—A review. Cogent Food & Agriculture. 2016;**2**:1267691. DOI: 10.1080/23311932.2016.1267691

[91] Boon MR, Bakker LE, Haks MC, Quinten E, Schaart G, Van Beek L, et al. Short-term high-fat diet increases macrophage markers in skeletal muscle accompanied by impaired insulin signalling in healthy male subjects. Clinical Science. 2015;**128**:143-151. DOI: 10.1042/CS20140179

[92] Patsouris D, Cao JJ, Vial G, Bravard A, Lefai E, Durand A, et al. Insulin resistance is associated with MCP1-mediated macrophage accumulation in skeletal muscle in mice and humans. PLoS One. 2014;**9**:110-653. DOI: 10.1371/journal.pone.0110653

[93] Fink LN, Costford SR, Lee YS, Jensen TE, Bilan PJ, Oberbach A, et al. Pro-inflammatory macrophages increase in skeletal muscle of high fat-fed mice and correlate with metabolic risk markers in humans. Obesity. 2014;**22**:747-757. DOI: 10.1002/oby.20615

[94] Khan IM, Perrard XY, Brunner G, Lui H, Sparks LM, Smith SR, et al. Intermuscular and perimuscular fat expansion in obesity correlates with skeletal muscle T cell and macrophage infiltration and insulin resistance. International Journal of Obesity. 2015;**39**:1607-1618. DOI: 10.1038/ijo.2015.104

[95] Wu H, Ballantyne CM. Skeletal muscle inflammation and insulin resistance in obesity. The Journal of Clinical Investigation. 2017;**127**:43-54. DOI: 10.1172/JCI88880

[96] Yaspelkis BB 3rd, Kvasha IA, Figueroa TY. High-fat feeding increases insulin receptor and IRS-1 coimmunoprecipitation with SOCS-3, IKKalpha/beta phosphorylation and decreases PI-3 kinase activity in muscle. American Journal of Physiology. Regulatory, Integrative and Comparative Physiology. 2009;**296**:1709-1715. DOI: 10.1152/ajpregu.00117.2009

[97] Todd JA, Walker NM, Cooper JD, Smyth DJ, Downes K, Plagnol V, et al. Robust associations of four new chromosome regions from genome-wide analyses of type 1 diabetes. Nature Genetics. 2007;**39**:857-864. DOI: 10.1038/ng2068

[98] Kaneto H, Nakatani Y, Kawamori D, Miyatsuka T, Matsuoka TA. Involvement of oxidative stress and the JNK pathway in glucose toxicity. Nature Medicine. 2004;**1**:165-174. DOI: 10.1900/RDS.2004.1.165

[99] Yuan M, Konstantopoulos N, Lee J, Hansen L, Li ZW, Karin M, et al. Reversal of obesity- and diet-induced insulin resistance with salicylates or targeted disruption of Ikkbeta. Science. 2001;**293**:1673-1677. DOI: 10.1126/science.1061620

[100] Hundal RS, Petersen KF, Mayerson AB, Randhawa PS, Inzucchi S, Shoelson SE, et al. Mechanism by which high-dose aspirin improves glucose metabolism in type 2 diabetes. The Journal of Clinical Investigation. 2002;**109**:1321-1326. DOI: 10.1172/JCI14955

[101] Kim JK, Kim YJ, Fillmore JJ, Chen Y, Moore I, Lee J, et al. Prevention of fat-induced insulin resistance by salicylate. The Journal of Clinical Investigation. 2001;**108**:37-46. DOI: 10.1172/JCI11559

[102] Hirosumi J, Tuncman G, Chang L, Görgün CZ, Uysal KT, Maeda K, et al. A central role for JNK in obesity and insulin resistance. Nature. 2002;**420**:333-336. DOI: 10.1038/nature01137

[103] Singh R, Wang Y, Xiang Y, Tanaka KE, Gaarde WA, Czaja MJ. Differential effects of JNK1 and JNK2 inhibition on murine steatohepatitis and insulin resistance. Hepatology. 2009;**49**:87-96. DOI: 10.1002/hep.22578

[104] Hardin BJ, Campbell KS, Smith JD, Arbogast S, Smith J, Moylan JS, et al. TNF-alpha acts via TNFR1 and muscle-derived oxidants to depress myofibrillar force in murine skeletal muscle. Journal of Applied Physiology. 2008;**104**:694-699. DOI: 10.1152/japplphysiol.00898.2007

[105] Tracey D, Klareskog L, Sasso EH, Salfeld JG, Tak PP. Tumor necrosis factor antagonist mechanisms of action: A comprehensive review. Pharmacology & Therapeutics. 2008;**117**:244-279. DOI: 10.1016/j.pharmthera.2007.10.001

[106] Aggarwal BB. Signalling pathways of the TNF superfamily: A double-edged sword. Nature Reviews. Immunology. 2003;**3**:745-775. DOI: 10.1038/nri1184

[107] Watanabe K, Iizuka T, Adeleke A, Pham L, Shlimon AE, Yasin M, et al. Involvement of toll-like receptor 4 in alveolar bone loss and glucose homeostasis in experimental periodontitis. Journal of Periodontology. 2011;**46**:21-30. DOI: 10.1111/j.1600-0765.2010.01304.x

[108] Liang H, Hussey SE, Sanchez-Avila A, Tantiwong P, Musi N. Effect of lipopolysaccharide on inflammation and insulin action in human muscle. PLoS One. 2013;**8**:e63983. DOI: 10.1371/journal.pone.0063983

[109] Goodman SC, Letra A, Dorn S, Araujo-Pires AC, Vieira AE, Chaves de Souza L, et al. Expression of heat shock proteins in periapical granulomas. Journal of Endodontia. 2014;**40**:830-836. DOI: 10.1016/j.joen.2013.10.021

[110] Nakhjavani M, Morteza A, Esteghamati A, Khalilzadeh O, Zandieh A, Safari R. Serum lipoprotein(a) levels are greater in female than male patients with type-2 diabetes. Lipids. 2011;**46**:349-356. DOI: 10.1007/s11745-010-3513-1

[111] Hunter-Lavin C, Hudson PR, Mukherjee S, Davies GK, Williams CP, Harvey JN, et al. Folate supplementation reduces serum hsp70 levels in patients with type 2 diabetes. Cell Stress & Chaperones. 2004;**9**:344-349. DOI: 10.1379/CSC-28R.1

[112] Stolarczyk E, Lord GM, Howard JK. The immune cell transcription factor T-bet: A novel metabolic regulator. Adipocytes. 2013;**3**:58-62. DOI: 10.4161/adip.26220

[113] Rocha VZ, Folco EJ, Sukhova G, Shimizu K, Gotsman I, Vernon AH, et al. Interferon-gamma, a Th1 cytokine, regulates fat inflammation: A role for adaptive immunity in obesity. Circulation Research. 2008;**103**:467-476. DOI: 10.1161/CIRCRESAHA.108.177105

[114] Wright BL, Nguyen N, Shim KP, Masterson JC, Jacobsen EA, Ochkur SI, et al. Increased GATA-3 and T-bet expression in eosinophilic esophagitis versus gastroesophageal reflux disease. The Journal of Allergy and Clinical Immunology. 2018;**141**:1919-1921. DOI: 10.1016/j.jaci.2017.12.993

[115] O'Rourke RW, White AE, Metcalf MD, Winters BR, Diggs BS, Zhu X, et al. Systemic inflammation and insulin sensitivity in obese IFN-γ knockout mice. Metabolism. 2012;**61**:1152-1161. DOI: 10.1016/j.metabol.2012.01.018

[116] Wada T, Hoshino M, Kimura Y, Ojima M, Nakano T, Koya D, et al.

Both type I and II IFN induce insulin resistance by inducing different isoforms of SOCS expression in 3T3-L1 adipocytes. American Journal of Physiology. Endocrinology and Metabolism. 2011;**300**:1112-1123. DOI: 10.1152/ajpendo.00370.2010

[117] Chang YH, Ho KT, Lu SH, Huang CN, Shiau MY. Regulation of glucose/lipid metabolism and insulin sensitivity by interleukin-4. International Journal of Obesity. 2012;**36**:993-998. DOI: 10.1038/ijo.2011.168

[118] Ricardo-Gonzalez RR, Red Eagle A, Odegaard JI, Jouihan H, Morel CR, Heredia JE, et al. IL-4/STAT6 immune axis regulates peripheral nutrient metabolism and insulin sensitivity. Proceedings of the National Academy of Sciences of the United States of America. 2010;**107**:22617-22622. DOI: 10.1073/pnas.1009152108

[119] Greiwe JS, Cheng B, Rubin DC, Yarasheski KE, Semenkovich CF. Resistance exercise decreases skeletal muscle tumor necrosis factor alpha in frail elderly humans. The FASEB Journal. 2001;**15**:475-482. DOI: 10.1096/fj.00-0274com

[120] Ferreira JP, Leal AMO, Vasilceac FA, Sartor CD, Sacco ICN, Soares AS, et al. Decreased muscle strength is associated with proinflammatory cytokines but not testosterone levels in men with diabetes. Brazilian Journal of Medical and Biological Research. 2018;**51**:7394. DOI: 10.1590/1414-431X20187394

Section 3

Muscle Wasting: Cancer-Induced

Chapter 6

Leucine and Its Importance for Cell Signalling Pathways in Cancer Cachexia-Induced Muscle Wasting

Andre Gustavo Oliveira, Bread Cruz,
Sarah Christine Pereira de Oliveira, Lais Rosa Viana,
Natalia Angelo Da Silva Miyaguti,
Luiz Alberto Ferreira Ramos, Rafael Rossi Valentim
and Maria Cristina Cintra Gomes-Marcondes

Abstract

The anabolic effects of a supplemented diet with branched-chain amino acids, especially leucine, on skeletal muscle wasting and as a co-adjuvant in cancer treatment have been well-studied. Leucine is a precursor of protein synthesis and acts as a nutritional signal, affecting multiple metabolic processes (e.g., satiety, thermogenesis, energy efficiency, and body composition). Previous studies related to nutritional therapy have mainly focused on myopenia, which is the loss of skeletal muscle mass in some pathologies, including cancer. Leucine plays a role in the maintenance and even increase of lean body mass in healthy individuals as well as the prevention of disease states that culminate in myopenia. Herein, we review the available data addressing the mechanisms by which leucine acts as a cellular signal, thereby stimulating muscle protein synthesis, leading to the inhibition of muscle catabolism, especially in an experimental model of cancer cachexia. We also show differences found in the metabolomic and proteomic analyses, including the use of leucine in maternal diets as a preventative for muscle wasting as supported by our experimental data.

Keywords: leucine, cell signalling, protein metabolism, protein synthesis, protein degradation, muscle wasting, experimental cachexia models

1. Introduction

Cancer remains an enigmatic pathology for some patient types and can also cause deleterious effects, e.g., in some cases ending in a cachexia state. Cancer cachexia is a complex syndrome that results from anorexia associated with glucose intolerance, depletion of body fat, and severe wasting of lean mass, which corresponds to the more significant proportion of metabolically active tissue—the muscle tissue. In particular, the loss of skeletal mass, which is referred to as myopenia in the pathological process, is clinically relevant as this process is directly related to the loss of muscle function in cancer patients. In every type of cachexia, the

IntechOpen

pathogenesis of muscle loss is complex and multifactorial. Due to the high energy expenditure produced by neoplastic cells, the patient presents inefficiency of energy production, as well as so-called futile energy processes, which is a major cause of muscle wasting. These energy expenses include high glucose production via gluconeogenesis from lactate or gluconeogenic precursors where there is excessive consumption associated with reduced the production of ATP. Then, the cancer patient loses weight involuntarily with severe loss of muscle mass due to increased protein degradation, which produces gluconeogenic amino acids. Moreover, this process also includes lipolysis, resulting in glycerol as a glucose precursor, inducing a spoliation cycle. All these points lead the cancer patient into fatigue and asthenia, thereby leading to a worse prognosis. Recently, the number of studies on new cancer treatment therapies has increased; most of these studies focus on the patient's responses to conventional treatments and improvement of survival and quality of life.

A novel therapeutic approach to cancer involves preserving, restoring, or even an epigenetic influence to maintain an adequate nutritional status for cancer patients, thereby slowing the onset of muscle mass wasting. In this context, nutritional supplementation has been identified as a potentially useful intervention. During protein synthesis, branched-chain amino acids (BCAAs), mainly leucine, act as precursors of the carbon skeleton and nitrogen. Also, leucine can primarily be oxidised in the muscle for energy supply and contribute nitrogen for the synthesis of other amino acids. Leucine also plays an essential role in cell signalling, stimulates protein synthesis, and modulates catabolism, mainly in skeletal muscle. In an experimental cachexia model that includes a leucine-rich diet (data in print), we report an improvement in functional muscle tests (verified by CatWalk test) and the influence of maternal leucine supplementation on the offspring's adulthood responses in the improving of the muscle tissue response. Since recent works have indicated that the most important goal during cancer progression is the maintenance of lean body mass and considering the key role of leucine in modulating skeletal muscle protein synthesis and degradation, our research group has been evaluating the effects of a leucine-rich diet in an experimental model of cancer cachexia. Herein, we summarise the findings of our group as well as others that show that a leucine-rich diet can ameliorate the prognosis, reduce the risk of death, and help to maintain the quality of life in cancer patients.

2. Cancer cachexia

Cachexia is a condition characterised by reduced food intake; involuntary and progressive weight loss; and intense catabolism of carbohydrates, lipids, and proteins [1], thereby resulting in intense deterioration of host tissues, severe weight loss, and adipose tissue and muscle mass wasting [2]. Weight loss and malnutrition are the most common characteristics observed in advanced cancer patients [3]. Cachexia is responsible for almost 30% of all cancer-related deaths and associated with significantly decreased physical activity and psychological burden [3]. Cachexia is also related to other pathophysiological changes, such as systemic inflammation, insulin resistance, and oxidative stress [4]. Several pro-inflammatory cytokines (e.g., IL-6 and TNF-α) and pro-cachectic factors (e.g., factor inducing proteolysis [PIF] and lipid mobilisation factor [LMF]), which are considered mediators of muscle wasting, act during the cachectic process [5, 6]. Considering the high prevalence (50–80%) of cachexia in advanced cancer, the investigation of the molecular process of cancer cachexia is important when considering the most efficient targets of treatment. The use of nutritional interventions to minimise the side effects of cancer is a novel and promising approach [7]. As such, supplementation with BCAA, especially leucine, has been shown to improve skeletal muscle mass because leucine plays an important role in skeletal muscle metabolism, regulates protein synthesis by stimulating the mTOR pathway, and inhibits the ubiquitin-proteasome pathway [8, 9].

3. Muscle protein synthesis and leucine

Muscle mass is a controlled balance of protein turnover by the cellular processes of protein synthesis and breakdown. In some pathological conditions, such as cancer cachexia, protein synthesis could also be compromised, which results in skeletal muscle atrophy and weakness [10, 11]. The regulation of protein turnover in skeletal muscle is a complex process, usually involving interactions between gene transcription, translation, and protein degradation. Stimulation and signalling processes are initiated by principal agents of these activities, such as anabolic hormones (e.g., insulin), growth factors, glucose, and amino acids. One such signalling pathway is triggered by insulin, which initiates protein synthesis after binding to its receptor, thereby activating several downstream components. The activated insulin receptor triggers the tyrosine phosphorylation of the insulin receptor substrate (IRS) 1 and 2, followed by activation of the phosphoinositol 3 kinase (PI3-kinase). Then, PI3-kinase activates phosphoinositide kinases-dependent 1 and 2 (PDK1/2) to phosphorylate the protein kinase Akt/PKB. The activated PKB phosphorylates tuberous sclerosis 2 (TSC2) inactivates the tuberous sclerosis complex 1 and 2 (TSC1/TSC2), which are no longer able to perform GTPase activity at Rheb (brain-enriched ras-homologue), allowing Rheb to release and activate the mechanistic target of rapamycin (mTOR), the key component of this machinery [12]. Also, the Akt substrate, PRAS40, when phosphorylated by PKB, loses its inhibitory effect over mTOR. In fact, mTOR acts as a sensor and integrator of diverse inputs, such as nutrients, growth factors, and energy status. mTOR, which consists of the mTORC1 and mTORC2 complexes, is a master regulator of protein synthesis and is essential for the maintenance of muscle mass and function [13]. Upon encountering anabolic factors, such as amino acids, mTORC1 is activated and signals to ribosomal protein S6 protein kinase 1 (S6K1) and eukaryotic initiation factor 4E binding protein 1 (4EBP1)—these are the best-known downstream effectors of mTOR signalling and control the protein synthesis pathway. Activated p70S6K subsequently leads to phosphorylation of the downstream target S6K1, which results in the translation of messenger RNA (mRNA) encoding for ribosomes and transcription factors [14]. In parallel, mTOR phosphorylates the dissociation of the 4E-BP1/eIF4E complex, releasing the eukaryotic initiation factor 4E (eIF4E), which subsequently binds to eIF4G, thus forming the eIF4F translation initiation complex, and allows the recruitment of the 40S ribosomal subunit to initiate protein translation [15].

Although the impairment of muscle protein synthesis in cachexia is not an obligate feature, many studies are working in strategies to improve the muscle mass and also the patient's muscle function which could imply in better prognosis and quality of life in those patients. Thereby, leucine together with valine and isoleucine, or even alone, can stimulate protein synthesis and act as cell signalling molecules in skeletal muscle by activating the mTOR pathway [16, 17]. Multiple studies have shown that leucine alone stimulates protein synthesis, mediating the translational control of protein synthesis in skeletal muscle independently of other BCAAs [18, 19]. For example, some studies have shown that the oral administration or infusion of leucine in adult humans or animals elevates muscular protein synthesis [20]. Moreover, leucine appears to have a much more potent anabolic effect (i.e., stimulating the mTOR pathway) than anabolic hormones, such as insulin. The administration of leucine after fasting or amino acid starvation stimulates protein synthesis and promotes the phosphorylation and activation of S6K1 via the rapamycin-sensitive mTOR in skeletal muscle [21, 22]. On the other hand, several studies have emphasised the specific contribution of cell membrane transport through the coupling of the amino acid transporter system 1 (LAT1 or SLC7A5), which carries leucine in exchange for glutamine [23]. The availability of amino acids (especially leucine and glutamine) is determined by its uptake by the cell, which appears to play an essential role in the entrance of Leu into the cell and the maintenance of a high intracellular concentration of glutamine [24]. Some evidence suggests that leucine uses the insulin signalling pathway, but the exact mechanism of triggering the mTOR complex remains under debate [25]. Nutritional supplementation with leucine stimulated the incorporation of phenylalanine in muscle in an experimental cachexia model, confirming an increased protein synthesis and also an increasing muscle mass [26, 27]. Subunits of the mTORC1 complex (i.e., Raptor and GβL) and substrates belonging to the downstream pathway (i.e., 4E-BPs, eIF4A, eIF4B, eIF4E, including S6K1), which represent the key points within the metabolism of proteins through mTOR, are highly increased in the muscle of Walker-256 tumour-bearing animals subjected to leucine nutritional supplementation [27–29]. *In vitro* cell culture studies have generated evidence relevant to the mechanism through which leucine affects mTOR [30, 31]. As such, leucine supplementation can stimulate protein synthesis and, consequently, might lead to a positive protein net balance, even within a high rate of protein degradation. In addition to increasing the protein synthesis in skeletal muscle, a leucine-rich diet has a protective effect in other tissues. In our previous works, we also observed improvement of protein synthesis in placenta tissue since leucine acted by improving the cell-signalling activity, thereby increasing placental protein synthesis and also reducing the placental proteolytic process [32, 33].

Interestingly, our previous works and other experimental studies have shown that leucine supplementation can work as an excellent nutritional strategy to treat or prevent muscle wasting in cancer cachexia. In this way, leucine also emerged as a potent stimulator of metabolism, leading to improvements in both oxidative metabolism and mitochondrial biogenesis [34]. Recently, our research group used metabolomic and proteomic analyses in an experimental cachexia model to better understand the benefits of leucine supplementation. Compared to a non-supplemented group, tumour-bearing rats under leucine supplementation showed metabolic pathways diverted to ketone bodies and butyrate metabolism [35]. Since an excess of leucine might provide ketone precursors being utilised by muscle tissue as energy sources, this likely diverted the metabolism to improve muscle protein synthesis [35]. The ketone bodies could provide additional energy to skeletal muscle and host tissues; this energy source is not available to the non-leucine-supplemented group. Besides acting as a fuel source to supply energy for the cellular activity of several tissues, ketone bodies, especially acetoacetate, can also promote muscle cell proliferation [36], probably accounting for the benefits of leucine nutritional supplementation [35]. We also made important findings as part of our proteomic analysis of the muscle tissue of tumour-bearing rats fed a leucine-rich diet (data in print). These results show a significant action of leucine on modulation of the mitochondrial membrane proteins involving the production of ATP, such as the ATP synthase complex family. Proteins associated with ATP synthase (e.g., F1F0 or Complex V) participate in the synthesis of ATP from ADP in the presence of a proton gradient across the mitochondrial membrane. One protein from this family that stood out in our studies is the ATP5a1 synthase subunit alpha. The tumour-bearing group showed a higher concentration of ATP5a1, which indicates a higher mitochondrial activity for the production of ATP, which is associated with a greater availability of glucose from the gluconeogenesis process. In contrast, the leucine tumour-bearing group showed lower muscle ATP5a1 content, likely indicating that the production of ATP must be derived from other metabolic processes. Therefore, the presence of tumour factors interferes in the cellular processes involved with obtaining energy. Since cell proliferation depends on the constant use of ATP for the duplication of all cellular machinery, the interference of tumour factors in the muscle mitochondria and the electron transport chain leads to less availability to energy for muscle cell activity. With leucine supplementation, a stimulating pathway occurs to obtain energy, thus contributing to the maintenance of adequate ATP supply and the ability maintain muscle activities. Since leucine is a ketogenic amino acid, its entire carbon skeleton is converted to keto acid or acetyl-CoA, which can be directed to participate in the Krebs cycle and beta-oxidation processes, both of which produce ATP as the final product. According to our results, the metabolic pathways in cachectic-tumour-bearing animals are related to ammonia recycling and the urea cycle, likely associated with protein degradation and directly associated with the futile cycle of energy production. In parallel, the metabolic activity of the leucine-tumour-bearing group was affected, which was related to ketone body and butyrate metabolism [35]. These points confirm the relationship to the increase in the muscle tissue's energy needs in tumour-bearing animals, which are minimised/modulated when the animal's diet is supplemented with leucine. Our proteomic results show that, in muscle tissue, mitochondria dysfunction occurs in the tumour-bearing host; however, under leucine supplementation, there are muscle mitochondrial biogenesis and activities improvements (data not published). Thus, we know that both insulin and leucine can independently affect the activation of mRNA translation and, consequently, the protein synthesis process [17, 37]. However, the real effect of amino acid signalling, especially leucine, on protein synthesis via the mTOR pathway remains complex and less understood, and there is a need for further studies, especially *in vivo* models. Moreover, as mentioned previously, our data for the CatWalk analysis showed an improvement in muscular functional activity; i.e., when rats with tumours were fed a leucine-rich diet, muscle function improved (data in print). Moreover, a leucine-supplemented maternal diet can influence and ameliorate the adult host response in tumour-bearing rats. In this way, an improved understanding of muscle protein synthesis and how leucine influences it is essential when developing new targets and strategies to restore muscle in muscle wasting diseases.

4. Muscle protein degradation and leucine

4.1 Skeletal muscle wasting in cachexia

Muscle homeostasis is important because muscle makes up a large part of the whole organism and performs many functions and activities. Moreover, as one of the main structures of the body, it is the most significant source of the protein turnover process. As noted above, muscle maintenance occurs by the intense activity of both protein synthesis and degradation [38]. Accordingly, proteolytic systems also play a key role in the regulation of cellular homeostasis and cell recycling, differentiation, cell cycle, abnormal protein degradation, and amino acid supply for gluconeogenesis [39]. Myofibrillar protein degradation is performed by the following four different pathways: ubiquitin proteasome system (UPS), autophagy, Ca^{2+}-dependent proteolysis, and caspase pathway [40]. The increase of protein catabolism in skeletal muscle contributes to a worse prognosis in cancer patients, especially those in a cachexia state, which is one of the most important causes of morbidity and mortality in these patients [41]. In cancer patients, the loss of either skeletal or cardiac muscle mass might lead to cardiac and respiratory failure, in addition to the fact that it decreases the host's response after conventional treatments, such as chemotherapy or radiotherapy [42]. Moreover, it is well-established that the ubiquitin-proteasome system is a very important pathway in skeletal muscle degradation during cancer cachexia [43]. Furthermore, multiple studies have identified released factors that contribute to an increase in muscle protein degradation during cancer. The main factors that lead to protein degradation during cancer cachexia syndrome produced by the host are tumour necrosis factor alpha (TNF-α), interleukin-6 (IL-6), interleukin-1 (IL-1), interferon gamma (IFN-γ) [44]. Meanwhile, the main factors produced by cancer cells are proteolysis-inducing factor (PIF), lipid-mobilising factor (LMF), and anaemia-inducing factor (AIS) [44].

Among the factors produced by a tumour, the proteolysis-inducing factor (PIF) has central importance. This protein, first described by Todorov et al. [6], is a 24 kDa glycoprotein isolated from the adenocarcinoma MAC16 tissue, an experimental model of cancer cachexia. Similarly, some studies found a PIF like those that were also verified in other cachexia models, such as in Walker-256 carcinosarcoma [31], in patients with gastrointestinal [45], pancreatic [46], and other types of cancer [47]. The injection of PIF in mice induces an intense loss of lean body mass, similar to that associated with MAC16 tumour growth [6]. After being synthesised and released by tumour cells, PIF reaches the bloodstream and binds to its cell membrane receptor in muscle cells [48], leading to activation of the ubiquitin-proteasome pathway and a decrease in protein synthesis by stimulating the double-stranded RNA-dependent protein kinase (PKR) [49]. The activated PKR leads to phosphorylation of eukaryotic translation initiation factor (eIF2α) and, consequently, inhibition of protein synthesis [50]. Many studies have shown that, unlike starvation, a decrease in food intake is not sufficient to cause muscle mass wasting in cancer patients, such as that which occurs during cancer cachexia [51].

As mentioned above, despite the fact that leucine stimulates protein synthesis, leucine and its metabolite β-hydroxy-β-methylbutyrate (HMB) can also decrease the rate of protein degradation apparently by reducing the expression of proteins from the ubiquitin-proteasome system [52, 53] and the other proteolytic pathways, i.e., mainly autophagy [54]. This characteristic makes leucine a great tool in cancer-induced cachexia therapy. In fact, leucine or HMB, i.e., alone [55] or in combination with other nutrients [56], can prevent the decrease of lean mass in cancer patients; this has been verified by our group in an experimental cachexia model [32, 35].

4.1.1 Proteolytic pathways and leucine

As noted above, the UPS is responsible for degrading proteins and might be responsible for up to 80% of proteolysis during skeletal muscle wasting [57, 58]. Since the UPS depends on linking the target protein to a ubiquitin tag and subsequent recognition and degradation by the proteasome core, leucine cell signalling can affect multiple steps. Ubiquitin conjugation to target proteins involves the action of a ubiquitin-activating enzyme (E1), which uses ATP to form thioester ubiquitin; conjugating to the ubiquitin-conjugating protein family of enzymes (E2), which in concert with ubiquitin protein ligase (E3), mediates the binding of the ubiquitin C terminal end to the targeted protein. The specificity of the substrate recognition is mainly dependent on E3 interaction with the targeted protein, giving relevance to this class of enzymes in studies of muscle atrophy affected by tumour evolution [59]. Our previous studies have shown that leucine supplementation can minimise the E2 activity in the muscle of Walker-256 tumour-bearing rats, suggesting a beneficial effect of this cell signal (data in print). Although approximately 1000 members of the E3 ligase family have been described, MuRF1 and

Atrogin have been reported to be specifically expressed and increased in skeletal muscle under many catabolic conditions [60, 61]. Interestingly, acute or chronical leucine supplementation prevented the upregulation of proteasomal proteolysis in fasted aged rats as compared to younger adult controls [62]. We have recently verified that ageing causes additional proteasomal activity in an experimental model of cancer-induced cachexia (unpublished data). Thus, leucine supplementation might be a valuable tool to counteract higher susceptibility to cachexia in senescence. Also, MuRF1, atrogin, and other E3 ligases, such as MUSA1 and SMART1, have been associated with enhanced proteolysis during muscle wasting [63]. Moreover, the degradation of ubiquitin-tagged proteins occurs in proteasome 26S, formed by regulatory (19S) and catalytic (20S) subunits. Interaction of the 19S subunit with ubiquitin drives the target protein to the core of the proteasome, a cylindrical protein complex formed by two external alpha rings (alpha 1–7) and two central beta rings (beta 1–7), in which beta 1, 2, and 5 present caspase-like, trypsin, and chymotrypsin protease activity, respectively [64]. Proteasome degradation results in 7–9 amino acid peptides, which are subsequently degraded by cytosolic proteases. During muscle wasting, the activity of chymotrypsin is increased, as is the expression of 19S, 11S, and 20S, all of which are modulated by the nutritional supplementation of leucine [27, 65, 66].

In addition to UPS in muscle wasting, autophagy is a degradation process led by lysosomes, and it manages the catabolism of long-life proteins, defective organelles, and protein aggregates. Three different autophagy pathways have been described, i.e., microautophagy, chaperone driven autophagy, and macroautophagy (herein referred as autophagy)—extensively reported as a key regulator of muscle mass. Autophagy involves complex protein machinery, including ATGs (autophagy-related genes); ultimately, autophagy leads to the formation of phagophores, i.e., the formation of autophagosomes by the elongation of the lipid membrane, which is followed by a fusion of the autophagosome to the lysosome, generating the autolysosome with many hydrolases and proteases (i.e., cathepsins). The first step of autophagy is the activation of ULK1, which, in turn, phosphorylates Beclin1, promoting its interaction with VPS34, VPS15, and ATG14. This complex activates VPS34, assembling the phagophore rich in PI3K class III enzyme to form PI3P, a signal to recruit other ATGs. The ubiquitin-like ATG5 brings together the final complex, i.e., ATG5/ATG12/ATG16, initiating and expanding the membrane extension of the phagophore. In parallel, the conjugated form of ATG8, also a homologue to LC3 in muscle, is tightly bound to the autophagosome membrane and later cleaved by ATG4, thereby converting LC3I to LC3II, which is necessary for the fusion of the autophagosome with the endocytic compartments, thus forming the lysosome. Inside the autolysosome, cargoes are degraded with cathepsins L and B being especially important for the degradation of myofibrils proteins [67]. Interestingly, the treatment of C2C12 myotubes with PIF-like increased the cathepsin B and chymotrypsin-like activity. The previous exposition of leucine PIF-like-treated myotubes prevented not only cathepsins and chymotrypsin enzymes activity but also proteasome activity [31], thereby highlighting another role for leucine in cancer-cachexia reversal. Indeed, the inhibition of cathepsin activity has been suggested as a useful approach to treat cancer cachexia [68].

Calcium-dependent proteolysis is composed by cysteine-proteases, which are dominated by calpains and the endogenous inhibitor calpastatin [69]. Among the 14 calpains described, striated muscle contains considerable amounts of μ-calpain, m-calpain and calpain 3, which are activated by the intracellular concentration of calcium [70]. Above a certain threshold, Ca^{2+} intracellular levels interact with the C-terminal domain of the calpain large subunit, thereby promoting N-terminal auto-cleavage and leading to maximal protease activity. Therefore, there is a correlation between calpain activity and protein turnover in muscle, thereby suggesting an important role in muscle mass maintenance and partially accounting for muscle wasting in some pathologies [71]. Cancer patients and tumour-bearing rats present higher calcium-dependent proteolysis, which is linked with increased calpain and decreased calpastatin protein content [72, 73]. Moreover, there is evidence that calpains, mainly μ-calpain, is localised to the Z-disk in the sarcomere, which is anchored by myofibrils such as p35, nebulin, troponin-T, α-activin, and desmine. Thus, Ca^{2+} dependent proteolysis seems essential to the initial disaggregation of the sarcomere structure, thereby releasing contractile proteins for further degradation by other proteolytic systems. Similarly, leucine supplementation improves muscle mass, thereby minimising the muscle wasting by inhibiting the calpain activity.

Other points of the proteolysis processes that should be mentioned include those related to other intracellular pathways that govern cancer cachexia. For example, AKT phosphorylation causes FoxOs inactivation and translocation from the nucleus to the cytosol [74]. FoxOs (Foxo1, 3 and 4) are transcription factors that regulate energy metabolism, the cell cycle, antioxidant defence, cell death, and longevity [75]. Repression of FoxOs by AKT is a key step in the anticatabolic action of insulin/IGF1 signalling [76]. Therefore, genetic ablation of FoxOs specifically in skeletal muscle reverses muscle atrophy caused by starvation and denervation, indicating that FoxOs are necessary to the expression of several atrophic genes, such as atrogin1, MuRF1, proteasome subunits, and lysosomal enzymes [75]. FoxO directly upregulated many proteasome subunits and E3 ligases in cancer-induced muscle wasting. Interestingly, FoxOs also increased autophagy in tumour-bearing mice [77], inducing the expression of such genes as Cathepsin-L and other genes related to the lysosomal/autophagy pathway (e.g., Gabarapl1 and Bnip3). Thus, FoxOs seem to mediate crosstalk between proteasomal and autophagy-dependent proteolysis in cancer-induced cachexia since the inhibition of FoxO3 or FoxO1 by RNA interference entirely prevents muscle loss [78]. Also, pro-inflammatory cytokines contribute to mass muscle decline under several conditions [79]. TNFα increases muscle protein degradation by activating the transcription factor NFκB [38]. The blockade of NFκB signalling in tumour-bearing mice partially attenuated cancer-induced muscle loss, thereby enhancing longevity [80]. Activation of NFκB enhanced atrophy by the transcription of MuRF1, ubiquitin, UbcH2 (E2), proteasome subunits, and autophagy-related genes [81]. Likewise, higher levels of myostatin and activin-A (i.e., members of the TGF-B family, share the receptor ACTRIIβ, and are known to regulate muscle mass) are related to muscle atrophy [82, 83]. Moreover, the inhibition of the bioactivity of activin-A and myostatin by inhibin and follistatin prevents muscle loss independent of tumour growth [84]. Additionally, several human tumour cell lines secrete considerable amounts of myostatin and activin-A [85], which are correlated with muscle strength loss [86]. Moreover, myostatin acting due to ACTRIIβ downstream effectors Smad2/3 activity also enhances skeletal muscle loss by phosphorylating the Smad2/3 and transcription of MuRF1, atrogin1, and autophagy induction [87], which corroborates our data, thereby highlighting the modulatory effect of leucine supplementation in C2C12 cells treated with PIF-like. Interestingly, myostatin effects might depend on the suppression of PI3K/Akt signalling [88], where we also find some beneficial effects of leucine supplementation, such as restoring the inhibitory effect of Akt and minimising proteolysis in PIF-like-treated C2C12 cells.

5. Myocardial muscle in cachexia and leucine

Recently, the number of studies addressing cancer and cardiac failure has increased. This is because cancer has significant effects on skeletal muscle, causing a catabolic state and resulting in widespread and progressive atrophy, including myocardial tissue. Studies have shown that cardiac atrophy can be a result of cancer evolution and its treatments [89–91]. These damages result in symptoms that can include breathlessness, lethargy, reduced exercise tolerance, congestive cardiac failure, and mortality [92]. Because the alterations in cardiac muscle structure and metabolism induced by cancer cachexia are poorly understood, cardiologists and oncologists are working together to explore models of care to improve outcomes. Some findings show that pancreatic, lung, and colorectal cancer patients have a reduced heart mass with a reduced left ventricular (LV) and wall thickness and are associated with smaller heart cell size and numbers and increased extracellular stroma surrounding the myocytes [92]. Indeed, this cardiac atrophy is part of a complex systemic metabolic syndrome caused by cancer damage, resulting in severe muscle wasting, including of the myocardium. Rodent models of cancer cachexia also show characteristics of cardiac atrophy, including decreased heart weight and LV mass; the thinning of septal, interventricular, and posterior walls; and chamber dilation as demonstrated by echocardiography [91, 93]. Cardiac atrophy in cancer cachexia is likely driven by cellular atrophy, including the activation of UPS [93] and the imbalance of protein turnover [91]. Cancer cachectic mice presented decreased cardiac contractile function and heart rate with concomitantly increased heart tissue fibrosis, which was associated with higher pro-inflammatory cytokine content and enhanced oxidative stress [94, 95]. Therefore, the use of an experimental cancer cachexia model allows us to evaluate how leucine supplementation counteracts cachexia damage in the heart. Recently, we reported relevant data related to the benefits of leucine supplementation for reverting/maintaining cardiac mass for both tumour-bearing rats fed a leucine-rich diet [91] and adult offspring whose mothers had been fed a leucine-rich diet (data not published). More interestingly, we observed improvement in enzyme activities related to the heart function via electrocardiography as a positive effect of leucine in tumour-bearing rats [91]. We know that leucine stimulates protein synthesis through activation of the mTOR pathway, thereby stimulating the intracellular signalling pathways that modulate cellular metabolism and apoptosis; this supports our data since the activation of mTOR is also essential for mediating physiologic cardiac hypertrophy and preventing cardiac dysfunction in the face of pressure overload [91, 96], thus supporting the cardioprotective effects of leucine over the cancer-cachexia-induced cardiac damages [91].

6. Leucine maternal diet influence over muscle wasting

Since the number of new cases of cancer is increasing every year, and most of these are attributed to environmental factors and lifestyle, prevention is a major target of cancer studies [97, 98]. In addition to maintaining a balanced diet throughout life, the influence of maternal diet on offspring's adulthood has been widely studied [99].

Among environmental factors, nutritional composition is the main factor in the modulation of gene expression, especially those related to metabolic pathways. The periods of gestation and lactation are considered crucial because the maternal diet exerts influence on the development and the plasticity of organs and tissues of the foetus/newborn [100]. The energy composition of foods, fatty acid composition, proteins, and micronutrients can modify several aspects of metabolism. Poor or imbalanced maternal diet, e.g., undernutrition, might contribute to a change in the metabolic programming of the offspring [101, 102], thereby increasing the risk of metabolic diseases (e.g., insulin resistance, obesity, type II diabetes), cardiovascular disorders (e.g., hypertension and atherosclerosis), hormonal imbalance, and even cancer incidence in the offspring [103]. Thus, some amount of prevention may be achieved through a balanced maternal diet, considering not only the proper nutrition but the nutritional scheme; this can be viewed as a long-term investment that benefits both the current generation and its descendants, i.e., one that can minimise the risk of diseases (e.g., cancer) in the mother and her adult children [104, 105]. Thereby, due to foetal and lifetime nutrition there are epigenetic modifications [101], which are stable heritable patterns of gene expression in the DNA and histone proteins [106], and may result in DNA methylation, histone modifications, and RNA interference. Global hypomethylation, global miRNA downregulation, specific promoter hypermethylation, histone deacetylation, and upregulation of epigenetic machinery have been reported in cancer [107, 108], which are related to epigenetic silencing of detoxifying enzymes, suppressor tumour genes, cellular cycle regulators, apoptosis inducers, and DNA repair genes [109].

Knowing the benefits of leucine for attenuating the cachectic state and preventive interventions [110–112], previous studies using animal models of cachexia indicate that maternal nutrition affects the development of cancer cachexia and its effects in offspring adulthood [113, 114]. In our previous work, a maternal diet supplemented with leucine had a positive impact on the adult offspring's ability to respond to a Walker-256 tumour, diminishing the cachexia index, modulating markers of hepatic damage functions, and increasing the antioxidant response of the liver [112]. In this same experimental procedure concerning muscle wasting, our unpublished data show that maternal leucine supplementation can minimise the cachectic index by preserving the skeletal muscle mass in adult offspring. These results are confirmed by the stimulatory effect on the expression of mTOR pathway proteins. We observed a significant activation of mTOR and p70S6K, which indicates the preservation of protein synthesis and a decrease in proteolysis (i.e., we also verified less tyrosine release in the perfusion procedure) in the gastrocnemius muscle of these adulthood Walker-256 tumour-bearing rats subjected to a leucine enriched maternal diet (data in print). In fact, these findings indicate that leucine supplementation can modulate the mTOR pathway, resulting in the preservation of protein synthesis (data in print) and protection against the damaging effects of the Walker-256 tumour. Thus, maternal leucine supplementation shows promise in terms of improving the response to cachexia, i.e., preventing muscle loss, and further studies are needed to better understand the epigenetic mechanisms involved in this modulation and how the parental influence can counteract the damages caused by cachexia.

7. Conclusion

In summary, in **Figure 1**, we present evidence demonstrating the key role of leucine in improving skeletal muscle protein synthesis and minimising muscle degradation; we also report some metabolomic and proteomic findings, which are ameliorated by a diet supplemented with leucine. Also, these data show the benefits of leucine supplementation in cases of cardiac cachexia and the potential that a leucine supplemented maternal diet has for improvement of the host response to cancer-cachexia-induced muscle damage. As found in our studies and reported by other research groups, leucine is a suitable co-adjuvant treatment in an experimental model of cancer cachexia. However, more translation human studies are needed to determine whether leucine supplementation is capable of modulating muscle mass in cancer cachexia patients.

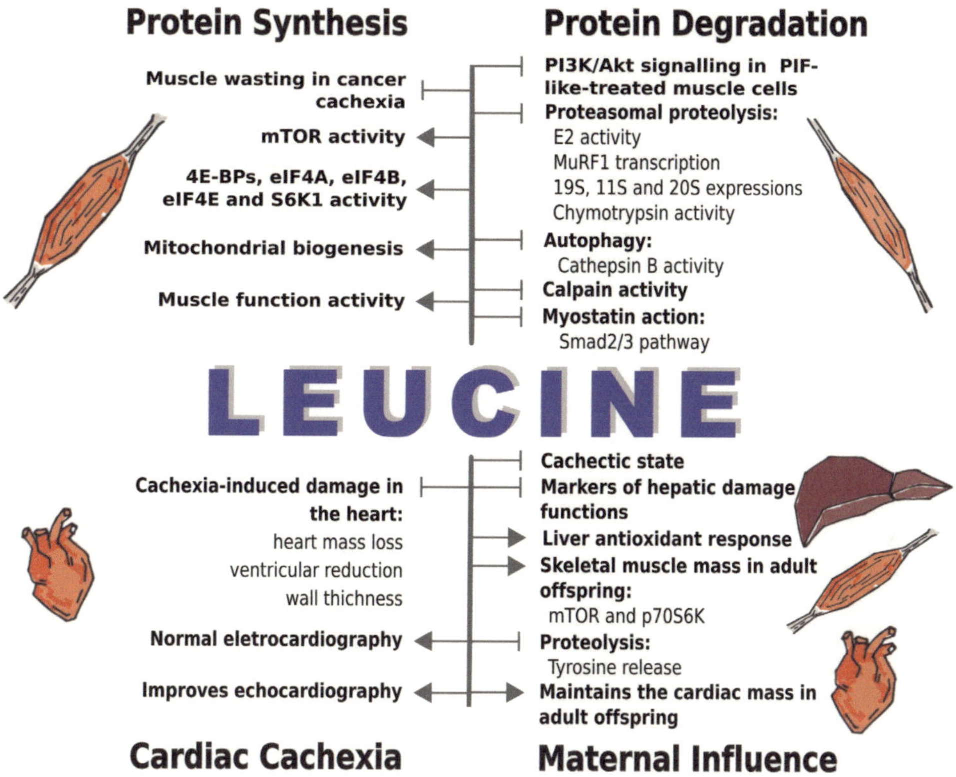

Figure 1.
Summary of the effects of leucine (and maternal diet) on protein synthesis, degradation, and cardiac cachexia. The arrows indicate the principal impact of leucine over the process, and the bar-headed lines show the inhibitory effect of leucine. All cited processes mentioned in the main text are summarised in these four blocks.

Acknowledgements

The authors are thankful for the financial support of Coordenação de Aperfeiçoamento de Pessoal de Nível Superior (CAPES), Conselho Nacional de Desenvolvimento Cientifico e Tecnológico (CNPq #302524/2016-9), and Fundação de Amparo à Pesquisa do Estado de São Paulo (FAPESP #2014/13334-7; #2015/06766-0; #2015/21890-0; #2017/02739-4; #2017/10809-2; #2017/23065-1). The authors gratefully thank Biol. R.W. Santos for technical support and A.C.G. Marcondes for the graphic art of tissue and organs. The main text has been edited by native speakers from American Manuscript Editors (Editing Certificate # 241-061-990-754-255).

Conflict of interest

The authors declare that there are no conflicts of interest.

Author details

Andre Gustavo Oliveira†, Bread Cruz†, Sarah Christine Pereira de Oliveira†, Lais Rosa Viana†, Natalia Angelo Da Silva Miyaguti†, Luiz Alberto Ferreira Ramos†, Rafael Rossi Valentim† and Maria Cristina Cintra Gomes-Marcondes*
Department of Structural and Functional Biology, Laboratory of Nutrition and Cancer, Institute of Biology, University of Campinas (UNICAMP), Campinas, Sao Paulo, Brazil

*Address all correspondence to: cintgoma@unicamp.br

† All authors are considered the first author, as all contributed equally to write, research, and develop data from experimental procedures for this work.

References

[1] Tisdale MJ. Are tumoral factors responsible for host tissue wasting in cancer cachexia? Future Oncology. 2010;**6**:503-513. DOI: 10.2217/fon.10.20

[2] Johns N, Stephens NA, Fearon KC. Muscle wasting in cancer. The International Journal of Biochemistry & Cell Biology. 2013;**45**:2215-2229. DOI: 10.1016/j.biocel.2013.05.032

[3] Skipworth RJE, Stewart GD, Dejong CHC, Preston T, Fearon KCH. Pathophysiology of cancer cachexia: Much more than host-tumour interaction? Clinical Nutrition. 2007;**26**: 667-676. DOI: 10.1016/j.clnu.2007.03.011

[4] Argilés JM, López-Soriano FJ, Busquets S, Argiles JM, Lopez-Soriano FJ, Busquets S. Mechanisms and treatment of cancer cachexia. Nutrition, Metabolism, and Cardiovascular Diseases. 2013;**23**(Suppl 1):S19-S24. DOI: 10.1016/j.numecd.2012.04.011

[5] Deans DAC, Wigmore SJ, Gilmour H, Paterson-Brown S, Ross JA, Fearon KCH. Elevated tumour interleukin-1beta is associated with systemic inflammation: A marker of reduced survival in gastro-oesophageal cancer. British Journal of Cancer. 2006; **95**:1568-1575. DOI: 10.1038/sj.bjc.6603446

[6] Todorov P, Cariuk P, McDevitt T, Coles B, Fearon K, Tisdale M. Characterization of a cancer cachectic factor. Nature. 1996;**379**:739-742. DOI: 10.1038/379739a0

[7] Maschke J, Kruk U, Kastrati K, Kleeberg J, Buchholz D, Erickson N, et al. Nutritional care of cancer patients: A survey on patients' needs and medical care in reality. International Journal of Clinical Oncology. 2017;**22**:200-206. DOI: 10.1007/s10147-016-1025-6

[8] Columbus DA, Fiorotto ML, Davis TA. Leucine is a major regulator of muscle protein synthesis in neonates. Amino Acids. 2015;**47**:259-270. DOI: 10.1007/s00726-014-1866-0

[9] Moro T, Ebert SM, Adams CM, Rasmussen BB. Amino acid sensing in skeletal muscle. Trends in Endocrinology and Metabolism. 2016; **27**:796-806. DOI: 10.1016/j.tem.2016.06.010

[10] Cohen S, Nathan JA, Goldberg AL. Muscle wasting in disease: Molecular mechanisms and promising therapies. Nature Reviews. Drug Discovery. 2015; **14**:58-74. DOI: 10.1038/nrd4467

[11] Argilés JM, Busquets S, Stemmler B, López-Soriano FJ. Cancer cachexia: Understanding the molecular basis. Nature Reviews. Cancer. 2014;**14**: 754-762. DOI: 10.1038/nrc3829

[12] Wang XJ, Yang X, Wang RX, Jiao HC, Zhao JP, Song ZG, et al. Leucine alleviates dexamethasone-induced suppression of muscle protein synthesis via synergy involvement of mTOR and AMPK pathways. Bioscience Reports. 2016;**36**(3):1-11. DOI: 10.1042/BSR20160096

[13] Mahoney SJ, Narayan S, Molz L, Berstler LA, Kang SA, Vlasuk GP, et al. A small molecule inhibitor of Rheb selectively targets mTORC1 signaling. Nature Communications. 2018;**9**(1): 548-559. DOI: 10.1038/s41467-018-03035-z

[14] Mieulet V, Roceri M, Espeillac C, Sotiropoulos A, Ohanna M, Oorschot V, et al. S6 kinase inactivation impairs growth and translational target phosphorylation in muscle cells maintaining proper regulation of protein turnover. American Journal of Physiology. 2007;**293**:C712-C722. DOI: 10.1152/ajpcell.00499.2006

[15] Gran P, Cameron-Smith D. The actions of exogenous leucine on mTOR signalling and amino acid transporters in human myotubes. BMC Physiology. 2011;**11**:10-19. DOI: 10.1186/1472-6793-11-10

[16] Hang YZ, Obayashi HK, Awatari KM, Ato JS, Ajotto GB, Itaura YK, et al. Effects of branched-chain amino acid supplementation on plasma concentrations of free amino acids, insulin, and energy substrates in young men. Journal of Nutritional Science and Vitaminology. 2011;**57**: 114-117

[17] Kimball SR, Jefferson LS. New functions for amino acids: Effects on gene transcription and translation. The American Journal of Clinical Nutrition. 2006;**83**:500S-507S

[18] Hernandez-García AD, Columbus DA, Manjarín R, Nguyen HV, Suryawan A, Orellana RA, et al. Leucine supplementation stimulates protein synthesis and reduces degradation signal activation in muscle of newborn pigs during acute endotoxemia. American Journal of Physiology. Endocrinology and Metabolism. 2016; **311**:E791-E801. DOI: 10.1152/ajpendo.00217.2016

[19] Kimball SR, Jefferson LS. Branched-chain amino acids: Metabolism, physiological function, and application signaling pathways and molecular mechanisms through which branched-chain amino acids mediate translational control of protein synthesis. The Journal of Nutrition. 2006;**136**:227-231

[20] Kitsy A, Carney S, Vivar JC, Knight MS, Pointer MA, Gwathmey JK, et al. Effects of leucine supplementation and serum withdrawal on branched-chain amino acid pathway gene and protein expression in mouse adipocytes. PLoS One. 2014;**9**:e102615. DOI: 10.1371/journal.pone.0102615

[21] Kimball SR, Shantz LM, Horetsky RL, Jefferson LS. Leucine regulates translation of specific mRNAs in L6 myoblasts through mTOR-mediated changes in availability of eIF4E and phosphorylation of ribosomal protein S6. The Journal of Biological Chemistry. 1999. DOI: 10.1074/jbc.274.17.11647

[22] Anthony JC, Yoshizawa F, Anthony TG, Vary TC, Jefferson LS, Kimball SR. Leucine stimulates translation initiation in skeletal muscle of postabsorptive rats via a rapamycin-sensitive pathway. The Journal of Nutrition. 2000;**130**:2413-2419

[23] Nicklin P, Bergman P, Zhang B, Triantafellow E, Wang H, Nyfeler B, et al. Bidirectional transport of amino acids regulates mTOR and autophagy. Cell. 2009;**136**:521-534. DOI: 10.1016/j.cell.2008.11.044

[24] Dodd KM, Tee AR. Leucine and mTORC1: A complex relationship. The American Journal of Clinical Nutrition. 2004;**79**:418-423. DOI: 10.1152/ajpendo.00525.2011

[25] Shimobayashi M, Hall MN. Making new contacts: The mTOR network in metabolism and signalling crosstalk. Nature Reviews. Molecular Cell Biology. 2014;**15**:155-162. DOI: 10.1038/nrm3757

[26] Ventrucci G, Mello MAR, Gomes-Marcondes MCC. Proteasome activity is altered in skeletal muscle tissue of tumour-bearing rats fed a leucine-rich diet. Endocrine-Related Cancer. 2004; **11**(4):887-895. DOI: 10.1677/erc.1.00828

[27] Cruz B, Oliveira A, Gomes-Marcondes MCC. L-leucine dietary supplementation modulates muscle protein degradation and increases pro-inflammatory cytokines in tumour-bearing rats. Cytokine. 2017; **96**:253-260. DOI: 10.1016/j.cyto.2017.04.019

[28] Soliman GA. The mammalian target of rapamycin signaling network and gene regulation. Current Opinion in Lipidology. 2005;**16**:317-323

[29] Ventrucci G, Mello MAR, Gomes-Marcondes MCC. Leucine-rich diet alters the eukaryotic translation initiation factors expression in skeletal muscle of tumour-bearing rats. BMC Cancer. 2007;**7**:42. DOI: 10.1186/1471-2407-7-42

[30] Gonçalves EM, Gomes-Marcondes MCC. Leucine affects the fibroblastic Vero cells stimulating the cell proliferation and modulating the proteolysis process. Amino Acids. 2010; **38**:145-153. DOI: 10.1007/s00726-008-0222-7

[31] Gonçalves EM, Salomão EM, Gomes-Marcondes MCC. Leucine modulates the effect of Walker factor, a proteolysis-inducing factor-like protein from Walker tumours, on gene expression and cellular activity in C2C12 myotubes. Cytokine. 2013;**64**:343-350. DOI: 10.1016/j.cyto.2013.05.018

[32] Viana LR, Gomes-Marcondes MCC. A leucine-rich diet modulates the tumor-induced down-regulation of the MAPK/ERK and PI3K/Akt/mTOR signaling pathways and maintains the expression of the ubiquitin-proteasome pathway in the placental tissue of NMRI mice. Biology of Reproduction. 2015; **92**(2):49-56. DOI: 10.1095/biolreprod.114.123307

[33] Cruz BLG, da Silva PC, Tomasin R, Oliveira AG, Viana LR, Salomao EM, et al. Dietary leucine supplementation minimises tumour-induced damage in placental tissues of pregnant, tumour-bearing rats. BMC Cancer. 2016;**16**:58. DOI: 10.1186/s12885-016-2103-x

[34] Vaughan RA, Garcia-Smith R, Gannon NP, Bisoffi M, Trujillo KA, Conn CA. Leucine treatment enhances oxidative capacity through complete carbohydrate oxidation and increased mitochondrial density in skeletal muscle cells. Amino Acids. 2013;**45**:901-911. DOI: 10.1007/s00726-013-1538-5

[35] Viana LR, Canevarolo R, Luiz ACP, Soares RF, Lubaczeuski C, Zeri AC de M, et al. Leucine-rich diet alters the ^{1}H-NMR based metabolomic profile without changing the Walker-256 tumour mass in rats. BMC Cancer. 2016; **16**:764. DOI: 10.1186/s12885-016-2811-2

[36] Zou X, Meng J, Li L, Han W, Li C, Zhong R, et al. Acetoacetate accelerates muscle regeneration and ameliorates muscular dystrophy in mice. The Journal of Biological Chemistry. 2016; **291**(5):2181-2195. DOI: 10.1074/jbc.M115.676510

[37] Proud CG. Role of mTOR signalling in the control of translation initiation and elongation by nutrients. Current Topics in Microbiology and Immunology. 2004;**279**:215-244

[38] Porporato PE. Understanding cachexia as a cancer metabolism syndrome. Oncogene. 2016;5:e200. DOI: 10.1038/oncsis.2016.3

[39] Koepp DM. Cell cycle regulation by protein degradation. Methods in Molecular Biology. 2014;**1170**:61-73. DOI: 10.1007/978-1-4939-0888-2_4

[40] Sandri M. Protein breakdown in muscle wasting: Role of autophagy-lysosome and ubiquitin-proteasome. The International Journal of Biochemistry & Cell Biology. 2013;**45**: 2121-2129. DOI: 10.1016/j.biocel.2013.04.023

[41] Del Fabbro E. Current and future care of patients with the cancer anorexia-cachexia syndrome. American Society of Clinical Oncology Educational Book. American Society of Clinical Oncology Meeting. 2015;**35**: e229-e237. DOI: 10.14694/EdBook_AM.2015.35.e229

[42] Aversa Z, Costelli P, Muscaritoli M. Cancer-induced muscle wasting: Latest findings in prevention and treatment. Therapeutic Advances in Medical Oncology. 2017;**9**:369-382. DOI: 10.1177/1758834017698643

[43] Khal J, Hine AV, Fearon KC, Dejong CH, Tisdale MJ. Increased expression of proteasome subunits in skeletal muscle of cancer patients with weight loss. The International Journal of Biochemistry & Cell Biology. 2005;**37**: 2196-2206. DOI: 10.1016/j.biocel.2004.10.017

[44] Argilés JM, Busquets S, Rodrigo Moore-Carrasco FJL-S. The role of cytokines in cancer cachexia. In: Mantovani G, AS D, Akio I, MJ E, Rossi FF, Daniele S, et al., editors. Cachexia Wasting A Mod. Approach. Milano: Springer; 2006. pp. 467-475. DOI: 10.1007/978-88-470-0552-5_43

[45] Cabal-Manzano R, Bhargava P, Torres-Duarte A, Marshall J, Bhargava P, Wainer IW. Proteolysis-inducing factor is expressed in tumours of patients with gastrointestinal cancers and correlates with weight loss. British Journal of Cancer. 2001;**84**:1599-1601. DOI: 10.1054/bjoc.2001.1830

[46] Wigmore SJ, Todorov PT, Barber MD, Ross JA, Tisdale MJ, Fearon KCH. Characteristics of patients with pancreatic cancer expressing a novel cancer cachectic factor. The British Journal of Surgery. 2000;**87**: 53-58. DOI: 10.1046/j.1365-2168.2000.01317.x

[47] Williams ML, Torres-Duarte A, Brant LJ, Bhargava P, Marshall J, Wainer IW. The relationship between a urinary cachectic factor and weight loss in advanced cancer patients. Cancer Investigation. 2004;**22**:866-870

[48] Todorov PT, Wyke SM, Tisdale MJ. Identification and characterization of a membrane receptor for proteolysis-inducing factor on skeletal muscle. Cancer Research. 2007;**67**:11419-11427. DOI: 10.1158/0008-5472.CAN-07-2602

[49] Eley HLL, Russell STT, Tisdale MJJ. Effect of branched-chain amino acids on muscle atrophy in cancer cachexia. The Biochemical Journal. 2007;**407**:113-120. DOI: 10.1042/BJ20070651

[50] Eley HL, Russell ST, Tisdale MJ. Mechanism of activation of dsRNA-dependent protein kinase (PKR) in muscle atrophy. Cellular Signalling. 2010;**22**:783-790. DOI: 10.1016/j.cellsig.2010.01.002

[51] Mendes MCS, Pimentel GD, Costa FO, Carvalheira JBC. Molecular and neuroendocrine mechanisms of cancer cachexia. The Journal of Endocrinology. 2015;**226**:R29-R43. DOI: 10.1530/JOE-15-0170

[52] Smith HJ, Mukerji P, Tisdale MJ. Attenuation of proteasome-induced proteolysis in skeletal muscle by {beta}-hydroxy-{beta}-methylbutyrate in cancer-induced muscle loss. Cancer Research. 2005;**65**:277-283

[53] Holeček M. Beta-hydroxy-beta-methylbutyrate supplementation and skeletal muscle in healthy and muscle-wasting conditions. Journal of Cachexia, Sarcopenia and Muscle. 2017;**8**:529-541. DOI: 10.1002/jcsm.12208

[54] Girón MD, Vílchez JD, Shreeram S, Salto R, Manzano M, Cabrera E, et al. β-Hydroxy-β-methylbutyrate (HMB) normalizes dexamethasone-induced autophagy-lysosomal pathway in skeletal muscle. PLoS One. 2015;**10**: e0117520. DOI: 10.1371/journal.pone.0117520

[55] Wilkinson DJ, Hossain T, Hill DS, Phillips BE, Crossland H, Williams J, et al. Effects of leucine and its metabolite β-hydroxy-β-methylbutyrate on human skeletal muscle protein metabolism. The Journal of Physiology.

2013;**591**:2911-2923. DOI: 10.1113/jphysiol.2013.253203

[56] May PE, Barber A, D'Olimpio JT, Hourihane A, Abumrad NN. Reversal of cancer-related wasting using oral supplementation with a combination of beta-hydroxy-beta-methylbutyrate, arginine and glutamine. American Journal of Surgery. 2002;**183**:471-479

[57] Tawa NE, Odessey R, Goldberg AL. Inhibitors of the proteasome reduce the accelerated proteolysis in atrophying rat skeletal muscles. The Journal of Clinical Investigation. 1997;**100**(1):197-203. DOI: 10.1172/JCI119513

[58] Senf SM, Dodd SL, Judge AR. FOXO signaling is required for disuse muscle atrophy and is directly regulated by Hsp70. American Journal of Physiology-Cell Physiology. 2010;**298**(1):C38-C45. DOI: 10.1152/ajpcell.00315.2009

[59] Oliveira AG, Gomes-Marcondes MCC. Metformin treatment modulates the tumour-induced wasting effects in muscle protein metabolism minimising the cachexia in tumour-bearing rats. BMC Cancer. 2016;**16**:418. DOI: 10.1186/s12885-016-2424-9

[60] Gomes MD, Lecker SH, Jagoe RT, Navon A, Goldberg AL. Atrogin-1, a muscle-specific F-box protein highly expressed during muscle atrophy. Proceedings of the National Academy of Sciences of the United States of America. 2001;**98**:14440-14445. DOI: 10.1073/pnas.251541198

[61] Bodine SC, Latres E, Baumhueter S, Lai VK, Nunez L, Clarke BA, et al. Identification of ubiquitin ligases required for skeletal muscle atrophy. Science. 2001;**294**:1704-1708. DOI: 10.1126/science.1065874

[62] Combaret L, Dardevet D, Rieu I, Pouch M-N, Béchet D, Taillandier D, et al. A leucine-supplemented diet restores the defective postprandial inhibition of proteasome-dependent proteolysis in aged rat skeletal muscle. The Journal of Physiology. 2005;**569**: 489-499. DOI: 10.1113/jphysiol.2005.098004

[63] Milan G, Romanello V, Pescatore F, Armani A, Paik J-H, Frasson L, et al. Regulation of autophagy and the ubiquitin–proteasome system by the FoxO transcriptional network during muscle atrophy. Nature Communications. 2015;**6**:6670-6683. DOI: 10.1038/ncomms7670

[64] Arendt CS, Hochstrasser M. Identification of the yeast 20S proteasome catalytic centers and subunit interactions required for active-site formation. Proceedings of the National Academy of Sciences of the United States of America. 1997;**94**: 7156-7161

[65] Ventrucci G, Mello MARAR, Gomes-Marcondes MCCCC. Proteasome activity is altered in skeletal muscle tissue of tumour-bearing rats a leucine-rich diet. Endocrine-Related Cancer. 2004;**11**:887-895. DOI: 10.1677/Erc.1.00828

[66] Cruz B, Gomes-Marcondes MCC. Leucine-rich diet supplementation modulates foetal muscle protein metabolism impaired by Walker-256 tumour. Reproductive Biology and Endocrinology. 2014;**12**:2. DOI: 10.1186/1477-7827-12-2

[67] Morris CA, Morris LD, Kennedy AR, Sweeney HL. Attenuation of skeletal muscle atrophy via protease inhibition. Journal of Applied Physiology. 2005;**99**:1719-1727. DOI: 10.1152/japplphysiol.01419.2004

[68] Sudhan DR, Siemann DW. Cathepsin L targeting in cancer treatment. Pharmacology & Therapeutics. 2015;**155**:105-116. DOI: 10.1016/j.pharmthera.2015.08.007

[69] Goll DE, Thompson VF, Li H, Wei W, Cong J. The calpain system. Physiological Reviews. 2003;**83**:731-801. DOI: 10.1152/physrev.00029.2002

[70] Goll DE, Neti G, Mares SW, Thompson VF. Myofibrillar protein turnover: The proteasome and the calpains. Journal of Animal Science. 2008;**86**:E19-E35. DOI: 10.2527/jas.2007-0395

[71] Bartoli M, Richard I. Calpains in muscle wasting. The International Journal of Biochemistry & Cell Biology. 2005;**37**:2115-2133. DOI: 10.1016/j.biocel.2004.12.012

[72] Costelli P, De Tullio R, Baccino FM, Melloni E. Activation of Ca^{2+}-dependent proteolysis in skeletal muscle and heart in cancer cachexia. British Journal of Cancer. 2001;**84**:946-950. DOI: 10.1054/bjoc.2001.1696

[73] Smith IJ, Aversa Z, Hasselgren P-O, Pacelli F, Rosa F, Doglietto GB, et al. CALPAIN activity is increased in skeletal muscle from gastric cancer patients with no or minimal weight loss. Muscle & Nerve. 2011;**43**:410-414. DOI: 10.1002/mus.21893

[74] Gan L, Zheng W, Chabot J-G, Unterman TG, Quirion R. Nuclear/cytoplasmic shuttling of the transcription factor FoxO1 is regulated by neurotrophic factors. Journal of Neurochemistry. 2005;**93**:1209-1219. DOI: 10.1111/j.1471-4159.2005.03108.x

[75] Link W, Fernandez-Marcos PJ. FOXO transcription factors at the interface of metabolism and cancer. International Journal of Cancer. 2017; **141**:2379-2391. DOI: 10.1002/ijc.30840

[76] O'Neill BT, Lee KY, Klaus K, Softic S, Krumpoch MT, Fentz J, et al. Insulin and IGF-1 receptors regulate FoxO-mediated signaling in muscle proteostasis. The Journal of Clinical Investigation. 2016;**126**:3433-3446. DOI: 10.1172/JCI86522

[77] Judge SM, Wu C-L, Beharry AW, Roberts BM, Ferreira LF, Kandarian SC, et al. Genome-wide identification of FoxO-dependent gene networks in skeletal muscle during C26 cancer cachexia. BMC Cancer. 2014;**14**:997. DOI: 10.1186/1471-2407-14-997

[78] Liu C-M, Yang Z, Liu C-W, Wang R, Tien P, Dale R, et al. Effect of RNA oligonucleotide targeting Foxo-1 on muscle growth in normal and cancer cachexia mice. Cancer Gene Therapy. 2007;**14**:945-952. DOI: 10.1038/sj.cgt.7701091

[79] de Matos-Neto EM, JDCC L, de Pereira WO, Figuerêdo RG, DMDR R, Radloff K, et al. Systemic inflammation in cachexia—Is tumor cytokine expression profile the culprit? Frontiers in Immunology. 2015;**6**:629. DOI: 10.3389/fimmu.2015.00629

[80] Macpherson PCD, Wang X, Goldman D. Myogenin regulates denervation-dependent muscle atrophy in mouse soleus muscle. Journal of Cellular Biochemistry. 2011;**112**: 2149-2159. DOI: 10.1002/jcb.23136

[81] Cornwell EW, Mirbod A, Wu C-L, Kandarian SC, Jackman RW. C26 cancer-induced muscle wasting is IKKβ-dependent and NF-kappaB-independent. PLoS One. 2014;**9**:e87776. DOI: 10.1371/journal.pone.0087776

[82] Zhou X, Wang JL, Lu J, Song Y, Kwak KS, Jiao Q, et al. Reversal of cancer cachexia and muscle wasting by ActRIIB antagonism leads to prolonged survival. Cell. 2010;**142**:531-543. DOI: 10.1016/j.cell.2010.07.011

[83] Li Q, Kumar R, Underwood K, O'Connor AE, Loveland KL, Seehra JS, et al. Prevention of cachexia-like

syndrome development and reduction of tumor progression in inhibin-deficient mice following administration of a chimeric activin receptor type II-murine Fc protein. Molecular Human Reproduction. 2007; **13**:675-683. DOI: 10.1093/molehr/gam055

[84] Chen JL, Walton KL, Hagg A, Colgan TD, Johnson K, Qian H, et al. Specific targeting of TGF-β family ligands demonstrates distinct roles in the regulation of muscle mass in health and disease. Proceedings of the National Academy of Sciences of the United States of America. 2017;**114**: E5266-E5275. DOI: 10.1073/pnas.1620013114

[85] Gold E, Marino FE, Harrison C, Makanji Y, Risbridger G. Activin-β(c) reduces reproductive tumour progression and abolishes cancer-associated cachexia in inhibin-deficient mice. The Journal of Pathology. 2013; **229**:599-607. DOI: 10.1002/path.4142

[86] Loumaye A, de Barsy M, Nachit M, Lause P, van Maanen A, Trefois P, et al. Circulating Activin A predicts survival in cancer patients. Journal of Cachexia, Sarcopenia and Muscle. 2017;**8**:768-777. DOI: 10.1002/jcsm.12209

[87] Sandri M. Protein breakdown in cancer cachexia. Seminars in Cell & Developmental Biology. 2016;**54**:11-19. DOI: 10.1016/j.semcdb.2015.11.002

[88] Seiliez I, Sabin N, Gabillard J-C. FoxO1 is not a key transcription factor in the regulation of myostatin (mstn-1a and mstn-1b) gene expression in trout myotubes. American Journal of Physiology. Regulatory, Integrative and Comparative Physiology. 2011;**301**:R97-R104. DOI: 10.1152/ajpregu.00828.2010

[89] Loncar G, Omersa D, Cvetinovic N, Arandjelovic A, Lainscak M. Emerging biomarkers in heart failure and cardiac cachexia. International Journal of Molecular Sciences. 2014;**15**: 23878-23896. DOI: 10.3390/ijms151223878

[90] Rajagopalan V, Zhao M, Reddy S, Fajardo G, Wang X, Dewey S, et al. Altered ubiquitin-proteasome signaling in right ventricular hypertrophy and failure. American Journal of Physiology. Heart and Circulatory Physiology. 2013; **305**:H551-H562. DOI: 10.1152/ajpheart.00771.2012

[91] Toneto AT, Ferreira Ramos LA, Salomão EM, Tomasin R, Aereas MA, Gomes-Marcondes MCC. Nutritional leucine supplementation attenuates cardiac failure in tumour-bearing cachectic animals. Journal of Cachexia, Sarcopenia and Muscle. 2016;7:577-586. DOI: 10.1002/jcsm.12100

[92] Murphy KT. The pathogenesis and treatment of cardiac atrophy in cancer cachexia. American Journal of Physiology. Heart and Circulatory Physiology. 2016;**310**:H466-H477. DOI: 10.1152/ajpheart.00720.2015

[93] Musolino V, Palus S, Tschirner A, Drescher C, Gliozzi M, Carresi C, et al. Megestrol acetate improves cardiac function in a model of cancer cachexia-induced cardiomyopathy by autophagic modulation. Journal of Cachexia, Sarcopenia and Muscle. 2016;7:555-566. DOI: 10.1002/jcsm.12116

[94] Tian M, Asp ML, Nishijima Y, Belury MA. Evidence for cardiac atrophic remodeling in cancer-induced cachexia in mice. International Journal of Oncology. 2011;**39**:1321-1326. DOI: 10.3892/ijo.2011.1150

[95] Springer J, Tschirner A, Haghikia A, von Haehling S, Lal H, Grzesiak A, et al. Prevention of liver cancer cachexia-induced cardiac wasting and heart failure. European Heart Journal. 2014;

35:932-941. DOI: 10.1093/eurheartj/eht302

[96] Huang Y, Zhou M, Sun H, Wang Y. Branched-chain amino acid metabolism in heart disease: An epiphenomenon or a real culprit? Cardiovascular Research. 2011;**90**:220-223. DOI: 10.1093/cvr/cvr070

[97] Bail J, Meneses K, Demark-Wahnefried W. Nutritional status and diet in cancer prevention. Seminars in Oncology Nursing. 2016;**32**:206-214. DOI: 10.1016/j.soncn.2016.05.004

[98] World Health Organization. Global Status Report on Noncommunicable Diseases 2010. World Health Organization; 2011. p. 176. DOI: ISBN: 978 92 4 156422 9

[99] Remely M, Stefanska B, Lovrecic L, Magnet U, Haslberger AG. Nutriepigenomics. Current Opinion in Clinical Nutrition and Metabolic Care. 2015;**18**:328-333. DOI: 10.1097/MCO.0000000000000180

[100] Eberle C, Ament C. Diabetic metabolic programming: Mechanisms altering the intrauterine milieu. ISRN Pediatrics. 2012;**2012**:1-11. DOI: 10.5402/2012/975685

[101] Vanden Berghe W. Epigenetic impact of dietary polyphenols in cancer chemoprevention: Lifelong remodeling of our epigenomes. Pharmacological Research. 2012;**65**:565-576. DOI: 10.1016/j.phrs.2012.03.007

[102] Mathias PCF, Elmhiri G, De Oliveira JC, Delayre-Orthez C, Barella LF, Tófolo LP, et al. Maternal diet, bioactive molecules, and exercising as reprogramming tools of metabolic programming. European Journal of Nutrition. 2014;**53**:711-722. DOI: 10.1007/s00394-014-0654-7

[103] Wu G, Bazer FW, Cudd TA, Meininger CJ, Spencer TE. Maternal nutrition and fetal development. The Journal of Nutrition. 2004:2169-2172

[104] Victora CG, Adair L, Fall C, Hallal PC, Martorell R, Richter L, et al. Maternal and child undernutrition: Consequences for adult health and human capital. Lancet. 2008;**371**:340-357. DOI: 10.1016/S0140-6736(07)61692-4

[105] Perera F, Herbstman J. Prenatal environmental exposures, epigenetics, and disease. Reproductive Toxicology. 2011;**31**:363-373. DOI: 10.1016/j.reprotox.2010.12.055

[106] Tammen SA, Friso S, Choi SW. Epigenetics: The link between nature and nurture. Molecular Aspects of Medicine. 2013;**34**:753-764. DOI: 10.1016/j.mam.2012.07.018

[107] Taby R, Issa JJ. Cancer epigenetics. CA: A Cancer Journal for Clinicians. 2010;**60**:376-392. DOI: 10.3322/caac.20085

[108] Esteller M. Cancer epigenomics: DNA methylomes and histone-modification maps. Nature Reviews. Genetics. 2007;**8**:286-298

[109] Gerhauser C. Cancer chemoprevention and nutri-epigenetics: State of the art and future challenges. In: Pezzuto MJ, Suh N, editors. Natural Products in Cancer Prevention and Therapy. Berlin, Heidelberg: Springer Berlin Heidelberg; 2012. pp. 73-132. DOI: 10.1007/128_2012_360

[110] Hunter DC, Weintraub M, Blackburn GL, Bistrian BR. Branched chain amino acids as the protein component of parenteral nutrition in cancer cachexia. The British Journal of Surgery. 1989;**76**:149-153

[111] Tayek JA, Bistrian BR, Hehir DJ, Martin R, Moldawer LL, Blackburn GL. Improved protein kinetics and albumin

synthesis. Cancer. 1986:147-157. DOI: 10.1002/1097-0142

[112] Miyaguti NA da S, de Oliveira SCP, Gomes-Marcondes MCC. Maternal nutritional supplementation with fish oil and/or leucine improves hepatic function and antioxidant defenses, and minimizes cachexia indexes in Walker-256 tumor-bearing rats offspring. Nutrition Research. 2018;**51**:29-39. DOI: 10.1016/j.nutres.2017.12.003

[113] Togni V, Ota CCC, Folador A, Júnior OT, Aikawa J, Yamazaki RK, et al. Cancer cachexia and tumor growth reduction in Walker 256 tumor-bearing rats supplemented with N-3 polyunsaturated fatty acids for one generation. Nutrition and Cancer. 2003; **46**:52-58. DOI: 10.1207/S15327914NC4601_07

[114] Folador A, Hirabara SM, Bonatto SJR, Aikawa J, Yamazaki RK, Curi R, et al. Effect of fish oil supplementation for 2 generations on changes in macrophage function induced by Walker 256 cancer cachexia in rats. International Journal of Cancer. 2007;**120**:344-350. DOI: 10.1002/ijc.22333

Chapter 7

Adipose Tissue Remodeling during Cancer Cachexia

Miguel Luiz Batista Júnior and Felipe Henriques

Abstract

Cancer-induced cachexia (CC), characterized by systemic inflammation, body weight loss, adipose tissue (AT) remodeling, and muscle wasting, is a malignant metabolic syndrome with an undefined etiology. There is a consensus that multiple factors contribute to cancer-induced AT remodeling, and longitudinal studies show that patients lose AT before they start losing muscle mass. In CC, AT remodeling occurs predominantly through adipocyte atrophy, impairment of fatty acid turnover, inflammation, rearrangement of extracellular matrix (ECM), and browning of AT. More recently, some studies have shown that AT is affected early in the course of cachexia. Additionally, studies using experimental models have consistently indicated that the alterations in adipocyte metabolism begin quite early, followed by the downregulation of adipogenic and thermogenic genes. These sets of changes, in addition to metabolites derived from this process, maybe the initial (sterile) trigger of the sequence of events that result in the remodeling and dysfunction of AT in cachexia. Therefore, the present chapter aims to describe state of the art related to the subject of interest by analyzing the primary studies that have addressed the possible interface between inflammation and morphofunctional alterations of AT, in addition to the possible repercussions of this process during the development of CC.

Keywords: adipose tissue, cachexia, inflammation, metabolism, adipogenesis, ECM, browning

1. Introduction

1.1 Etiology of cachexia syndrome and adipose tissue remodeling

Cachexia is a common occurrence in the advanced stages of cancer and contributes to reduce the quality of life and life expectancy of patients [1, 2]. Cancer-related cachexia (CC) is a significant cause of morbidity and mortality, affecting more than 80% of individuals with advanced cancer and accounting for more than 20% of deaths [3–5]. Notably, the severity of cachexia does not correlate with tumor size [6]. Although there are descriptions of cachectic individuals dating back to more than 2000 years ago, increased attention has been focused on this syndrome in the recent decades. Nevertheless, its etiology remains unknown, and there is no treatment that is able to revert this condition [7, 8].

Cachexia is understood as a complex metabolic syndrome associated with underlying diseases and is characterized by a reduction of muscle mass and a depletion of fat stores [9–11]. Thus, the main clinical symptoms of CC are body weight loss in adults (corrected for water retention) and impaired (substandard) growth in children (after exclusion of endocrine disorders) [12]. Anorexia, inflammation, insulin resistance, and increased degradation of muscle proteins are frequently associated with cachexia [1, 2]. Although anorexia exhibits different characteristics when compared with starvation, muscle mass loss due to aging (sarcopenia), primary depression, malnutrition, and hyperthyroidism is correlated with increased morbidity associated with asthenia and metabolic disorders [3, 13].

Weight loss is the foremost independent predictor of mortality in cancer patients [8, 12, 14], beginning with the loss of both a fat mass (adipose tissue—AT) and the lean body mass (skeletal muscle tissue). Over the last few years, the former has often been proposed to proceed more rapidly in the patient than the latter [11, 15, 16]. By the late 1980s and early 1990s, cachexia was being approached from a different perspective, leading to a new conception according to which it is considered a chronic inflammatory syndrome. It is currently believed that factors produced by both the tumor and the host cause anorexia and the metabolic abnormalities that result in cachexia [3, 17, 18]. Based on the knowledge obtained regarding cachexia and its complexity, the latest international consensus defined standard diagnostic criteria for the disease [1]. According to this consensus, the condition is categorized as pre-cachexia (early stage), cachexia, or refractory cachexia (late stage) based on the severity of the following parameters: (1) the reduction of total body mass; (2) the presence of metabolic disorders; (3) anorexia; and (4) systemic inflammation, as illustrated in **Figure 1A**.

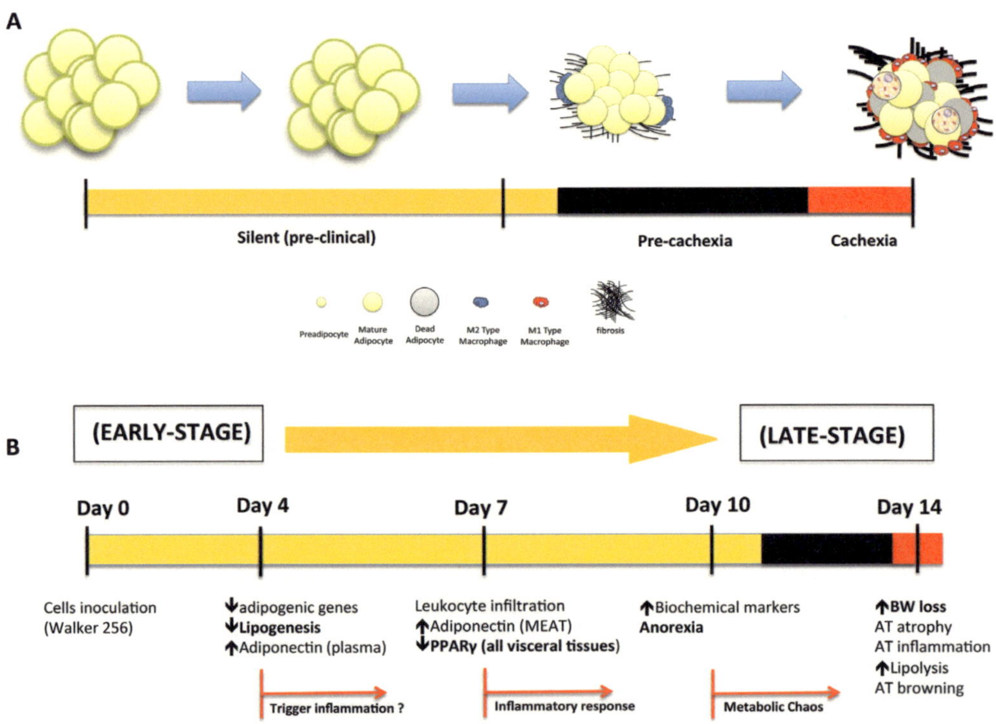

Figure 1.
Model of cachexia development from a translational point of view. (A) Morphofunctional changes in adipose tissue described in cancer cachexia patients. These alterations are associated with the stages of syndrome development, according to [1]. (B) Compilation of main metabolic and inflammatory changes described in the experimental model of cachexia induced by Walker 256 tumor.

An accurate understanding of the fundamental mechanisms that underlie CC is essential for the development of new pharmacological and nutritional therapies. In this way, several studies [19–23] have suggested that AT is the target of local and systemic factors derived from the host and by the tumor, including pro-cachectic factors [tumor necrosis factor α (TNF-α); interleukin (IL) 6; IL-1ß; IL-8; interferon-γ (INF-γ); ciliary neurotrophic factor; and leukemia inhibitory factor], anti-cachectic factors [soluble TNF-α receptor (sTNFR); soluble IL-6 receptor (sIL-6R); IL-1 receptor antagonist (IL-1RA); IL-4; IL-10; and IL-15], and tumor products [proteolysis-inducing factor (PIF); lipid-mobilizing factor (LMF); zinc-α 2-glycoprotein (ZAG); toxohormone-L; and anemia-inducing substance (AIS)]. Such factors are involved in the etiology and progression of cachexia [8, 14, 24–29]. Upon positing the inflammatory model as the hypothesis to be tested, some studies have recently demonstrated the relevance of subcutaneous adipose tissue (scAT) as both an important source of inflammatory mediators (particularly IL-6 and adiponectin) and an important source of biomarkers for cachexia (clinical progression), as adiponectin expression exhibits a correlation with the magnitude of the total body mass reduction [20]. More recently, a consistent modification consisting of inflammatory cell presence and fibrosis in scAT induced by cachexia was demonstrated in gastrointestinal cancer patients [30, 31]. The fibrosis was characterized by the presence of "crown-like structures" composed of CD68 positive AT macrophages (ATMϕs) surrounding adipocytes, and increased CD3 Ly, both of which were more evident in the fibrotic areas. In addition, some of these changes were already present in the cancer group, suggesting that AT inflammation may occur at an early stage of cachexia, even before the detection of pre-cachexic clinical features. Thus, alterations in AT inflammation seem to play a crucial role in the changes resulting in fat mass reduction, in addition to other morphofunctional alterations related to this tissue [19, 21]. Moreover, these changes appear to start quite early, long before any local tissue and/or systemic (circulatory system) changes can be detected.

Nevertheless, most studies on this topic have been restricted to assessing inflammation from the systemic point of view, only investigating plasma parameters, while neglecting tissue inflammation and, particularly, the events that precede the appearance of these alterations, although they may significantly contribute to disorders that result in AT remodeling, such as impairment of lipid metabolism, tissue cells turnover and inflammation, fibrosis, and subsequent systemic inflammation [2, 28]. Additionally, considering the important relationship among infiltrating cells (inflammatory mediators), the regulation of adipocyte metabolism, and the consequent remodeling of the AT, few or no studies have investigated this relationship and its role in the various stages of cachexia to our knowledge.

1.2 General features of adipose tissue remodeling

In general, AT remodeling in CC comprises a set of morphostructural modifications characterized by adipocyte atrophy, a result of an unbalance of lipids turnover, main due to increased lipolysis [10, 32]; impairment of adipocyte cellular turnover (adipogenesis) [18, 19]; enhanced inflammatory signaling [21, 30, 32]; modification of extracellular matrix (ECM) component generally resulting in fibrosis [30, 31]; and browning phenotype related to increase thermogenic effect [16, 33–35].

1.2.1 Adipose tissue atrophy is involved in the etiology of cachexia

The basic structure of the various types of AT includes mature adipocytes, stromal vascular cells (mesenchymal precursor cells, preadipocytes, fibroblasts, and immune system cells), blood vessels, lymph nodes, and nerves [36]. Fat cells (adipocytes) are the main cell type in AT, while the presence of other cell types varies as a function of the tissue location [mesenteric (meAT), epididymal (epiAT), retroperitoneal (rpAT), or scAT], animal species, and disease (e.g., obesity and cachexia) [37]. Additionally, the importance of AT in the control of adiposity is well established, as is the role of the adipokines (leptin, adiponectin, and resistin, among others) released by the adipose tissue [38]. While alterations in the development and metabolism of adipocytes have been implicated in the pathogenesis of human immunodeficiency virus (HIV)-related lipodystrophy [39], very little is known about the molecular mechanisms involved in the occurrence of lipodystrophy associated with other conditions, such as cachexia.

As mentioned above, in the course of cachexia, the observed body weight loss predominantly results from a reduction in the fat (AT) and lean (skeletal muscle) mass [10, 40]. More recently, AT loss was found to precede the reduction of the lean body mass [11, 15], and thus, a more accurate understanding of this process began to emerge. Several factors have been suggested as the cause of the changes that lead to a reduction of AT mass, including (1)increased lipid mobilization due to increased lipolysis in adipocytes [9, 10]; (2) reduced lipogenesis, resulting from decreased lipoprotein lipase (LPL) activity [17]; and (3) impaired adipocyte turnover, most likely as a function of an adipogenesis-apoptosis imbalance in AT [41].

Taking the factors that are most likely to be involved in the observed fat mass loss and the relevance of AT in cachexia into consideration, the recent studies have utilized an experimental animal model that allows temporal assessment of the main variables that are potentially involved in cachexia-induced AT disorders, with an emphasis on the parameters related to adipogenesis, metabolism, and inflammation [18, 23, 27, 32], as shown in **Figure 1A**. These data demonstrate that these alterations start early, at 4 days after cachexia induction by inoculation of Walker 256 tumor cells in Wistar rats [19, 32]. It is worth noting that this is the period that precedes the appearance of the classic symptoms of cachexia, such as dyslipidemia, and a reduction of the total body weight, as well as inflammatory alterations in the AT in these animals, which begin to predominate starting on Day 7. In Lewis lung carcinoma (LLC) tumor-bearing mice, upregulation of genes related to lipid turnover and adipose browning is evident even before the detection of the body weight loss of the animals [16, 33]. Similarly, using several experimental models of cachexia (syngeneic and genetic), K5-SOS mice showed reduction in fat mass and spleen enlargement in the pre-cachexic period, that is, before the detection of body weight loss and the development of cachexia [33]. Taking the aforementioned findings into consideration, the temporal characterization of some of the alterations that occur in different AT depots in the course of cachexia has effectively pointed to actual pathways and mediators that might be involved in the earliest changes, in addition to serving as biomarkers for the clinical progression of cachexia.

As mentioned above, accentuated reduction of fat mass is a significant clinical sign of CC. Although it has not yet been well established, these alterations depend on the location of the adipose tissue involved (e.g., visceral or subcutaneous). Taking this into account, it was recently shown that, considering results found in the experimental animal model of cachexia (Walker 256 tumors cells-induced), visceral fat depots were affected in different ways. Thus, epiAT and meAT [25] exhibited higher reductions in relative weight (percentage of total body mass), respectively, while rpAT did not show any change [42]. Furthermore, in cachectic patients, assessment of the adipose tissue area by means of computed tomography in humans showed that visceral AT was reduced in cachectic individuals with gastrointestinal carcinosarcoma when compared with controls [43]. On the other hand, in individuals with gastrointestinal cancer, scAT seems to be more affected when compared with visceral AT (epiAT and meAT), at least considering tissue inflammation parameters [44].

In fact, as described above, AT atrophy is a well-characterized clinical variable in cachexia syndrome. In addition, this tissue is affected early before the appearance of classic signs of cachexia. In this sense, the main morphological alteration observed as a result of AT atrophy is the alterations in the area and perimeter of adipocytes in both animal models [18, 42] and CC patients [20, 30]. Still in this context, depot-specific changes in adipocyte ultrastructure [42] were also described.

1.2.2 Metabolic features of adipocytes in the course of cachexia

The body weight loss in CC has been thought to be a result of profound changes in metabolic pathways of tissues and organs, which cannot be solely explained by enhanced energy expenditure or malnutrition [45]. In this regard, the role of early AT dysfunction seems to have gained importance in the onset and progression of many alterations induced by the syndrome. Different mechanistic possibilities have been proposed to explain the changes in AT in cachexia, such as increased lipolytic activity, decreased the activity of LPL, reduced de novo lipogenesis, and, consequently, decreased lipid triacylglycerol (TG) deposition [32, 45–47]. Adipocyte lipid turnover [i.e., the balance between incorporation and removal of TG into adipocytes, in which lipolysis (hydrolysis of intracellular TGs)] and is the most important factor for lipid removal [48]. In CC, human and animal models [11] have shown an increased rate of lipid mobilization, and longitudinal studies have shown that patients with CC lose AT mass before wasting of the muscle mass can be detected [15]. In addition, an accelerated rate of AT loss is believed to be associated with shorter survival time during cancer progression [49].

Most of the volume (>90%) of a white adipocyte is represented by a single fat droplet, which is mainly composed of TG. During periods of stress and/or nutrient deprivation, such as in metabolic disorders, the adipocytes activate mechanisms that lead to lipolysis, with a consequent release of non-esterified fatty acids (NEFAs) and glycerol originating in the TG stored in these cells. NEFAs are immediately released into the bloodstream and subsequently serve as a substrate for energy production of muscle tissues, or alternatively, they are taken up into the liver, where they are oxidized, esterified, or transformed into ketone bodies [50]. The reactions that result in NEFA release are mainly catalyzed by two enzymes: adipocyte triglyceride lipase (ATGL), which catalyzes the first step of the pathway, resulting in the formation of diacylglycerol, and hormone-sensitive lipase (HSL), which is responsible for additional hydrolysis and catalyzes the reactions that culminate in NEFA and glycerol release [10].

Among the mechanisms that may be involved in the metabolic disorders that cause fat mass reduction in cachexia, increased lipolysis appears to be the most evident and is being described as an increasing frequency [9, 11]. Das and colleagues found that in ATGL and HSL knockout mice, there was a greater resistance to the development of tumor-induced cachexia, which was more evident in ATGL-deficient animals. These authors also observed a positive correlation between ATGL activity and the severity of cachexia. Even more interestingly, they only detected a reduction of lean body mass in animals that exhibited severe cachexia. This phenomenon followed the events that led to a reduction of fat body mass, most noticeably involving an increase of TG hydrolysis in AT. The results of this study corroborate findings previously reported in individuals with cancer-related cachexia [15], thus increasing the consistency of the evidence demonstrating the role of AT as a target tissue in cachexia. Nevertheless, neither the exact time when these changes occur in the course of cachexia nor the various affected depots have yet been adequately described.

To elucidate these aspects, a study was conducted in which the lipolytic capacity of isolated adipocytes was assessed at 4 and 14 days after the inoculation of tumor cells. Two particular visceral fat stores were selected (meAT and rpAT) based on the results of previous studies that demonstrated their relevance for the development of cachexia. Day 4 was selected to perform the analysis, also based on previous results [19], which demonstrated downregulation of the genes involved in adipocyte metabolism, while no changes were found in the assessed morphological and inflammatory parameters. Interestingly, the tumor-bearing animals exhibited a reduction in basal lipolysis 4 days after inoculation, while on Day 14, the cachectic animals exhibited a considerable increase in the basal lipolysis rate (**Figure 1B**). These findings corroborate with the results of other studies showing increased lipolysis in the subcutaneous AT of cachectic patients [9, 51].

In this regard, another important aspect was a deregulation of lipolysis (in vitro) revealed a distinct profile, depending on the degree of disease progression. In the first, the basal rate of lipolysis was reduced and was accompanied by increased p-HSL (Ser565) expression, which is regulated by AMPK activation and, consequently, inhibits HSL activity. Patients with CC show reduced spontaneous basal lipolysis with elevated *ex vivo* responses to catecholamine and natriuretic peptides [40, 52]. In this aspect, an elegant study showed that the lack of AMPK activity is a common feature of adipose tissue dysfunction in cachectic mice and is triggered, at least partially, through the aberrant induction of *Cidea* and the subsequent degradation of AMPK in this tissue [53]. The authors suggest that treatment of cachectic animals with a peptide specifically targeting the white adipose tissue AMPK-CIDEA interaction prevents AT loss under CC conditions.

1.2.3 Downregulation of adipogenic genes in cancer cachexia

Adipogenesis may be defined as the process of the differentiation of precursor cells (preadipocytes) into new adipose cells (adipocytes) that are able to store TG and synthesize and secrete various proteins called adipokines [54]. In fact, impaired adipogenesis might contribute to the development of obesity-related metabolic disorders, such as peripheral insulin resistance, hyperlipidemia, and type 2 diabetes [55, 56]. The process of adipocyte differentiation involves the activation of a cascade of transcription factors that coordinate the expression of genes that are responsible for adipocyte function [55, 57]. The initial events include transient increases in CCAAT/enhancer-binding proteins beta and delta (C/EBPβ and δ), which allow preadipocytes to be distinguished from non-adipogenic precursor cells and subsequently activate the expression of the peroxisome proliferator-activated receptor gamma-2 (*Pparγ-2*), which in turn stimulates the expression of *C/ebpα*, which exerts a synergic effect with PPARγ-2 on the control of terminal differentiation [54, 57]. Local and endocrine factors might regulate adipogenesis through the modulation of these transcriptional events [58].

Adipocyte maturation is accompanied by intracellular lipid accumulation, which is mainly mediated by sterol regulatory element-binding protein-1C (SREBP-1C). In addition to activating the expression of *Pparγ*-2, SREBP-1C also activates the lipogenic pathway by stimulating the expression of the genes that encode the main enzymes of that pathway, such as ATP-citrate lyase (*Acly*), acetyl-CoA carboxylase (*Acaca*), fatty acid synthase (*Fasn*), and stearoyl-CoA desaturase-1 (*Scd-1*) [18, 57]. In addition to its direct participation in adipocyte differentiation, PPARγ-2 also plays an important role in the transcriptional regulation of genes associated with the lipogenic pathway by inducing the transcription of genes that encode adipocyte fatty acid-binding protein (*aP2*), LPL, fatty acid transport proteins (FATPs), and fatty acid-binding proteins (FABPs), among others. Activation of genes associated with glucose transport, such as *Glut4*, and thermogenesis, such as the mitochondrial uncoupling proteins (*Ucp2* and *Ucp3*), has also been described [18].

In AT, a balance between the growth/differentiation (adipogenesis) and death of its cells (generally by apoptosis) regulates the cellular turnover [41, 59]. In this aspect, the impairment of adipogenesis in the course of cachexia has been recently addressed. Some studies have elucidated the adipogenic marker profile during the development of cachexia syndrome [19, 25]. However, few studies have addressed the apoptotic processes and/or AT turnover during cachexia [43]. It has been known that adipogenic genes are downregulated in CC in epiAT [18] and rpAT [19]. On the other hand, scAT apoptosis did not change in cancer patients [43]. Therefore, considering that AT depots respond heterogeneously to CC and several metabolic and inflammatory pathways are involved in AT remodeling, there is no consensus if such effects induced by cachexia would be a result of secreted products directly by the tumor and tumor-host relationship. Thus, a recent study has analyzed in vitro adipogenesis in a co-culture system to mimic the effects of CC on adipocytes [60]. Co-culture of LLC cells promoted a decreased volume of the lipid droplets in 3 T3-L1 cells, compromising its maturation process (adipogenesis) in vitro. This result was followed by the downregulation of adipogenic and lipolytic gene expression, increased in apoptosis markers and proinflammatory cytokines secretion by both tumor cells and adipocytes. In this sense, these data suggest that the presence of the tumor cells was able to inhibit the adipocytes' maturation, which was associated with the increased levels of inflammatory cytokines.

In this way, the findings generated in the experimental model have demonstrated to be adequate for investigating cachexia-induced alterations of adipogenesis and point to the need to widen the scope of the assessed genes as well as the pathways and regulatory factors involved. Besides, modifications in adipogenesis appear to precede the appearance of the classic signs of cachexia as well as the signs of tissue inflammation in AT, such as increases in infiltrated macrophages and the production of inflammatory cytokines [21]. Thus, the factors that "silence" the genes involved in the differentiation of preadipocytes, and, consequently, in the maintenance of adipose cells turnover in AT might play a central role in the genesis of the damaging changes (metabolism and function) that occur in the adipose tissue of individuals with cachexia.

1.2.4 Adipose tissue inflammatory profile in the course of cachexia

According to the abovementioned findings from animal models and cancer cachexia patients, a metabolic and morphological dysfunction that results in the AT remodeling occurs during the development of CC [2]. More recently, another relevant aspect of cachexia-induced AT remodeling is the establishment of AT inflammation, which is characterized by increased recruitment of ATMϕs, including activated M1 (pro-inflammatory) and M2 (anti-inflammatory) macrophages [32]. In this way, an inflammatory profile predominates the end-stage of cachexia, particularly in visceral AT, as most of the characteristic changes of this syndrome are already established at this time [19, 27]. More recently, the immune cell infiltration profile was analyzed in greater detail, which was found to be characterized by an increase in M1-polarized macrophages [20]. In that study, the profile of chemokines specific to polymorphonuclear cells was also investigated, in addition to the presence of neutrophils in the various AT depots. The results revealed an increase in the chemokines CCL3 and CXCL2 at 7 days after tumor cell inoculation. The presence of CD11b-positive cells, which were tested to detect the presence of neutrophils, was observed in the same period. Taking the temporal changes identified in the cachexia model into consideration, as a function of the assessed parameters, the results indicate that inflammation starts on Day 7 and is established by Day 14, a period during which a series of disorders characteristic of cachexia become evident (e.g., a reduction of the total body and fat mass, dyslipidemia, and hypoglycemia). In this regard, even more recently, these findings were presented in greater detail in cachexia induced by LLC cells, showing ATMϕs polarization tends to be directed to M1 phenotype [61, 62].

Despite the increasing perception of the importance of the relationship between inflammation and CC, and systemic inflammation in particular, there is still no consensus regarding its source and also the role of inflammation of TA in the establishment and development of cachexia, among cancer patients in particular [9, 43]. Limitations in experimental designs, the selection of control groups, and the techniques used to analyze markers of inflammation have most likely been responsible for preventing a more precise investigation of the presence of inflammation in AT. Addressing this question, a study has recently demonstrated increased gene expression of phenotypic markers of ATMϕs and inflammatory cytokines, such as IL-6 and TNF-α, in cachectic patients with gastrointestinal cancer [20]. Interestingly, increased gene levels of IL-6 were positively correlated to increase plasmatic levels of this cytokine, indicating that in cachectic patients, scAT may be an important source of inflammatory mediators. Even more recently, the same group revealed an increase in ATMϕs forming crown-like structures in the same AT depot from cachectic patients [44], which is a characteristic finding in fat tissue in experimental animal models of obesity and in obese patients [63]. In addition, an increase in chemoattractant for ATMϕs gene expression in scAT, such as *Ccl2*, was detected only in cancer patients without cachexia, showing no changes in cachectic ones. However, despite the relevance of local inflammation, in AT in particular, the mechanisms responsible for both cachexia and inflammation remain to be elucidated. The characterization and understanding of the process of inflammation in cachexia are also relevant to establish whether it is secondary to or the "trigger" for the development of cachexia syndrome.

1.2.5 Alterations in the extracellular matrix of adipose tissue

Extracellular matrix (ECM) remodeling is the result of the processes of matrix synthesis and degradation during which specific proteins are deposited, such as tenascin and fibrin, and occurs under both physiological (e.g., tissue repair) and pathological conditions (e.g., inflammation) [64]. The ECM consists of a complex network of multifunctional and structural molecules, including various collagen isoforms, adhesive glycoproteins, and proteoglycans. This network provides support to cells and to the signaling pathways that control their migration, proliferation, and differentiation. Also, the ECM might serve as a reservoir of cytokines and other growth factors, which are released into the system in variable amounts depending on the pathological condition.

ECM remodeling plays a central role in the differentiation of adipocytes. Although the corresponding molecular mechanisms are only partially understood, ECM remodeling occurs concomitantly with the activation and/or repression of a transcriptional "network" involved in adipogenesis, which may be activated or repressed due to extracellular stimuli [65]. In the 3 T3-L1 mouse cell line, the differentiation of adipocytes is associated with a reduction in the fibronectin-rich matrix and basal lamina formation [66, 67]. Silencing of the pericellular collagen membrane type-1 matrix metalloproteinase (*Mt1-Mmp*) gene in the course of the mouse development results in the formation of a rigid collagen fibril chain and changes *in vivo* adipogenesis [68]. This condition is an example of the relationship between structural changes in ECM and adipocyte differentiation.

ECM remodeling plays a central role in the differentiation of adipocytes. Although the similar molecular mechanisms are only partially understood, ECM remodeling occurs concomitantly with the activation and/or repression of a transcriptional "network" involved in adipogenesis, which may be activated or repressed due to extracellular stimuli [69]. While some such alterations, such as changes in collagen content deposition, an increase in the number of infiltrated cells and insulin resistance, are also present in cancer-related cachexia, very little is known regarding the possible relationship between these cell types (e.g. fibroblasts, pre-adipocytes, immune cells, and others) and the processes that lead to ECM remodeling.

In MAC16 tumor-bearing mice, a cancer cachexia animal model, shrunken adipocytes and increased collagen-fibril content in AT were reported [18]. In CC patients, our group recently showed that the total type I collagen content of the scAT is rearranged in cachectic individuals with gastrointestinal cancer, which is associated with an increase in macrophage and lymphocyte contents. Interestingly, the total collagen content exhibited discrete changes in cancer patients without cachexia, but the expression of the *Ccl2* gene was found to be increased [20]. Another exciting finding demonstrates that ECM remodeling of AT in cancer cachexia results in augmented collagen fiber content. Excessive synthesis of mature elastic fibers accompanies such morphological scenery, besides strong labeling for collagen type I (COL1) and III (COL3) in the AT from cachectic patients [30, 31]. Besides, the presence of fibrosis was also associated with an increased number of myofibroblasts and an activated TGFβ/SMAD pathway in the subcutaneous AT of gastrointestinal cancer cachectic patients [31, 70].

These findings indicate that the morphological changes that lead to AT remodeling in CC patients are evident, albeit discretely, before the onset of the earliest characteristic symptoms of cachexia (**Figure 1**). However, to the best of our knowledge, no study has yet investigated the causes and repercussions of fat remodeling in the course of cachexia in full detail.

1.2.6 Browning of adipose tissue induced by cancer cachexia

Within the set of morphofunctional changes that result in the AT remodeling, it has recently been shown that cachexia induces browning of AT in addition to changes in immune-modulatory activity. In this scenario, chronic inflammation and β-adrenergic activation of thermogenesis functionally cooperate in the pathogenesis of cachexia [16, 33, 62, 71]. In general, browning of AT has been described as responsible for the increase in total caloric expenditure [72], and the induction of browning has therapeutic potential in promoting the reduction of body fat [16, 33, 34, 73]. However, this fact refers to conditions of diseases characterized by the presence of metabolic disorders, usually associated with high caloric intake, overweight, and/or obesity [74].

In this sense, in CC, the presence of the browning phenotype has shown to be detected very early [16] in different experimental models [33, 34, 35]. Also, it has demonstrated a significant role in altering the metabolism of this tissue because this process is related to the increase in energy expenditure and mobilization of fatty acids [17] by adipocytes. Another interesting new fact was that, in this same study, AT from cachectic mice showed upregulation of particular genes for brown adipocytes when compared to samples of brown adipose tissue. This fact is not usual because the one would expect an increase in genes specific to beige adipocytes. Also, regarding thermogenesis, the rectal temperature was reduced when the main clinical signs of the diseases were already established (refractory cachexia). Interestingly, this hypothermic phenotype has previously been described in the Walker 256 tumor-induced cachexia model, in the final stage of cachexia [75]. On the other hand, additional studies are needed to clarify the immuno-metabolic changes resulting in thermogenic adjustments induced by the syndrome, as well as the particular clinical consequences.

In this same study [33], the browning phenotype has also been described in samples of visceral adipose tissue from cancer cachexia patients. However, there is a need for analysis in a larger cohort and additional characterizations about the possible physiological repercussions for these patients. Also, there is still a need to characterize the real contribution of AT browning to overall energy expenditure during cancer cachexia. In this sense, an elegant study has evaluated, in several experimental models, that although the studies above detected mild induction of *Ucp1* mRNA levels in tumor-exposed AT, such changes appear to be discrete in thermogenic terms [53]. In this scenario, the overall effect of AT UCP1-dependent thermogenesis on systemic energy homeostasis may not be the principal actor during cancer cachexia.

2. Concluding remarks

In summary, several studies have shown that AT is significantly affected during the development of cachexia. The main alterations related to metabolic disorders, particularly those involving early adipocyte lipid turnover dysfunction of AT, increases in immune cell infiltration followed by increased local production of inflammatory mediators and remodeling of ECM components. More recently, some studies have shown that cachexia-induced browning of AT is a characteristic phenotype that arises from alterations that result in the AT remodeling, although its function is still not well characterized. Nevertheless, studies using those experimental models have consistently indicated that the modifications in the adipocyte metabolism begin quite early, and the metabolites derived from this process may be the initial (sterile) trigger of the sequence of events that result in the remodeling and consequent dysfunction of AT in cachexia. Finally, a deeper understanding of the initial stimulus that triggers AT dysfunction, in particular, inflammation and remodeling, needs to be further studied because evidence indicates that AT dysfunction plays a significant role in cachexia and may be a potential modulator of the process that could be explored therapeutically.

Acknowledgements

We would like to thank all the members of Laboratory of Adipose Tissue Biology for helpful discussions and critical reading of the chapter. We would also like to thank Alexander H. Bedard for the revision of the chapter as a native of the English language. This study was supported by São Paulo Research Foundation (FAPESP) Grants: 2010/51078-1, 2015/19259-0, and CNPq 311966/2015-2 to MLB Jr. The contents of this chapter are solely the responsibility of the authors and do not necessarily represent the official views of FAPESP and CNPq.

Conflict of interest

The authors declare no conflicts of interest.

Appendices and nomenclature

ECM	extracellular matrix
ATGL	adipose triglycerides lipase
HSL	hormone-sensitive lipase
CC	cancer-related cachexia
AT	adipose tissue
sTNFR	soluble tumor necrosis factor-α receptor
sIL-6R	soluble IL-6 receptor
IL-1RA	IL-1 receptor antagonist
IL	interleukin
PIF	proteolysis-inducing factor
LMF	lipid-mobilizing factor
ZAG	zinc-α2-glycoprotein
AIS	anemia-inducing substance
scAT	subcutaneous adipose tissue
CD	cluster of differentiation
ATMϕs	adipose tissue macrophage
CD3	cluster of differentiation 3
Ly	lymphocytes
HIV	human immunodeficiency virus
LPL	lipoprotein lipase
LLC	Lewis lung carcinoma
K5-SOS-F	keratinocyte-specific expression of an HA tagged dominant form of the human SOS1
TG	triacylglycerol
NEFAs	non-esterified fatty acids AMPK
AMPK	5′ adenosine monophosphate-activated protein
CIDEA	cell death-inducing DFFA-like effector a
C/EBP	CCAAT/enhancer-binding proteins
PPARγ	peroxisome proliferator-activated receptor gamma
SREBP-1C	sterol regulatory element-binding protein-1C
ACC	acetyl-CoA carboxylase
FAS	fatty acid synthase
SCD-1	stearoyl-CoA desaturase-1
GPAT	glycerol-3-phosphate acetyltransferase
aP2	adipocyte fatty acid-binding protein
FATPs	fatty acid transport proteins
FABPs	fatty acid-binding proteins
GLUT-4	glucose transport 4
UCP	mitochondrial uncoupling proteins
3 T3-L1	embryo fibroblast cells with a continuous substrain (L1) of 3T3 (Swiss albino) developed through clonal isolation
M1	polarized macrophages 1
M2	polarized macrophages 2
CCL	chemokine (C-C motif) ligand
CXCL	chemokine (C-X-C motif) ligand

MT1-MMP	collagen membrane type-1 matrix metalloproteinase
MAC16	murine adenocarcinoma 16
CCR2 (MCP-1)	C-C chemokine receptor type 2
COL1	collagen type I
COL3	collagen type III
TGFβ	transforming growth factor beta
SMAD	small worm phenotype mothers against decapentaplegic

Author details

Miguel Luiz Batista Júnior[1]* and Felipe Henriques[2]

1 Laboratory of Adipose Tissue Biology, Integrated Group of Biotechnology, University of Mogi das Cruzes, Mogi das Cruzes, Brazil

2 Program in Molecular Medicine, University of Massachusetts Medical School, Worcester, MA, USA

*Address all correspondence to: migueljr4@me.com

References

[1] Fearon K et al. Definition and classification of cancer cachexia: An international consensus. The Lancet Oncology. 2011;**12**(5):489-495

[2] Baracos VE et al. Cancer-associated cachexia. Nature Reviews Disease Primers. 2018;**4**:17105

[3] Bruera E, Sweeney C. Palliative care models: International perspective. Journal of Palliative Medicine. 2002; **5**(2):319-327

[4] Gordon JN, Green SR, Goggin PM. Cancer cachexia. QJM. 2005;**98**(11): 779-788

[5] Bing C, Trayhurn P. New insights into adipose tissue atrophy in cancer cachexia. Proceedings of the Nutrition Society. 2009;**68**:385-392

[6] Fearon KCH, Moses AGW. Cancer cachexia. International Journal of Cardiology. 2002;**85**(1):73-81

[7] von Haehling S et al. Cachexia: A therapeutic approach beyond cytokine antagonism. International Journal of Cardiology. 2002;**85**(1):173-183

[8] Argiles JM et al. The role of cytokines in cancer cachexia. Current Opinion in Supportive and Palliative Care. 2009; **3**(4):263-268

[9] Agustsson T et al. Mechanism of increased lipolysis in cancer cachexia. Cancer Research. 2007;**67**(11): 5531-5537

[10] Arner P. Medicine lipases in cachexia. Science. 2011;**333**(6039): 163-164

[11] Das SK et al. Adipose triglyceride lipase contributes to cancer-associated cachexia. Science. 2011;**333**(6039): 233-238

[12] Evans WJ et al. Cachexia: A new definition. Clinical Nutrition. 2008; **27**(6):793-799

[13] Inui A. Cancer anorexia-cachexia syndrome: Current issues in research and management. CA: A Cancer Journal for Clinicians. 2002;**52**(2):72-91

[14] Argilés JM, Alvarez B, López-Soriano FJ. The metabolic basis of cancer cachexia. Medicinal Research Reviews. 1997;**17**(5):477-498

[15] Fouladiun M et al. Body composition and time course changes in regional distribution of fat and lean tissue in unselected cancer patients on palliative care—correlations with food intake, metabolism, exercise capacity, and hormones. Cancer. 2005;**103**(10): 2189-2198

[16] Kir S et al. Tumour-derived PTH-related protein triggers adipose tissue browning and cancer cachexia. Nature. 2014;**513**(7516):100-104

[17] Tisdale MJ. Mechanisms of cancer cachexia. Physiological Reviews. 2009; **89**(2):381-410

[18] Bing C et al. Adipose atrophy in cancer cachexia: Morphologic and molecular analysis of adipose tissue in tumour-bearing mice. British Journal of Cancer. 2006;**95**(8):1028-1037

[19] Batista ML et al. Heterogeneous time-dependent response of adipose tissue during the development of cancer cachexia. The Journal of Endocrinology. 2012;**215**(3):363-373

[20] Batista ML et al. Adipose tissue-derived factors as potential biomarkers in cachectic cancer patients. Cytokine. 2013;**61**(2):532-539

[21] Batista ML et al. Adipose tissue inflammation and cancer cachexia:

Possible role of nuclear transcription factors. Cytokine. 2012;**57**(1):9-16

[22] Tsoli M, Robertson G. Cancer cachexia: Malignant inflammation, tumorkines, and metabolic mayhem. Trends in Endocrinology and Metabolism. 2013;**24**(4):174-183

[23] Tsoli M et al. Depletion of white adipose tissue in cancer cachexia syndrome is associated with inflammatory signaling and disrupted circadian regulation. PLoS One. 2014; **9**(3):e92966

[24] Argilés JM, López-Soriano FJ, Busquets S. Novel approaches to the treatment of cachexia. Drug Discovery Today. 2008;**13**(1-2):73-78

[25] Bao Y et al. Zinc-[alpha]2-glycoprotein, a lipid mobilizing factor, is expressed and secreted by human (SGBS) adipocytes. FEBS Letters. 2005; **579**(1):41-47

[26] Laviano A et al. Neural control of the anorexia-cachexia syndrome. American Journal of Physiology. Endocrinology and Metabolism. 2008; **295**(5):E1000-E1008

[27] Machado AP, Costa Rosa LFPB, Seelaender MCL. Adipose tissue in Walker 256 tumour-induced cachexia: Possible association between decreased leptin concentration and mononuclear cell infiltration. Cell and Tissue Research. 2004;**318**(3):503-514

[28] Vegiopoulos A, Rohm M, Herzig S. Adipose tissue: Between the extremes. The EMBO Journal. 2017;**36**(14): 1999-2017

[29] Shyh-Chang N. Metabolic changes during cancer cachexia pathogenesis. Advances in Experimental Medicine and Biology. 2017;**1026**:233-249

[30] Batista ML Jr et al. Cachexia-associated adipose tissue morphological rearrangement in gastrointestinal cancer patients. Journal of Cachexia Sarcopenia Muscle. 2016;**7**:37-47

[31] Alves MJ et al. Adipose tissue fibrosis in human cancer cachexia: The role of TGFβ pathway. BMC Cancer. 2017;**17**(1):190

[32] Henriques FS et al. Early suppression of adipocyte lipid turnover induces immunometabolic modulation in cancer cachexia syndrome. The FASEB Journal. 2017;**31**(5):1976-1986

[33] Petruzzelli M et al. A switch from white to brown fat increases energy expenditure in cancer-associated cachexia. Cell Metabolism. 2014;**20**(3): 433-447

[34] Han J et al. Interleukin-6 induces fat loss in cancer cachexia by promoting white adipose tissue lipolysis and browning. Lipids in Health and Disease. 2018;**17**(1):14

[35] Elattar S, Dimri M, Satyanarayana A. The tumor secretory factor ZAG promotes white adipose tissue browning and energy wasting. The FASEB Journal. 2018;**32**(1 Supplement)

[36] Pond CM. Adipose tissue and the immune system. Prostaglandins, Leukotrienes and Essential Fatty Acids. 2005;**73**(1):17-30

[37] Hausman DB et al. The biology of white adipocyte proliferation. Obesity Reviews. 2001;**2**(4):239-254

[38] Trayhurn P, Bing C, Wood IS. Adipose tissue and adipokines—energy regulation from the human perspective. The Journal of Nutrition. 2006;**136**(7): 1935S-11939S

[39] Capeau J et al. Diseases of adipose tissue: Genetic and acquired lipodystrophies. Biochemical Society Transactions. 2005;**33**(Pt 5): 1073-1077

[40] Dahlman I et al. Adipose tissue pathways involved in weight loss of cancer cachexia. British Journal of Cancer. 2010;**102**(10):1541-1548

[41] Arner P, Spalding KL. Fat cell turnover in humans. Biochemical and Biophysical Research Communications. 2010;**396**(1):101-104

[42] Bertevello PS, Seelaender MCL. Heterogeneous response of adipose tissue to cancer cachexia. Brazilian Journal of Medical and Biological Research. 2001;**34**:1161-1167

[43] Mikael R et al. Lipolysis—not inflammation, cell death, or lipogenesis—is involved in adipose tissue loss in cancer cachexia. Cancer. 2008;**113**(7):1695-1704

[44] Seelaender M, Batista ML. Hetrogeneus response of rat and human adipose tissue in cachexia. In: 5th cachexia conference. Barcelona: Spain: Society on Cachexia and Wasting Disorders (SCWD). 2009

[45] Argilés JM et al. Cancer cachexia: Understanding the molecular basis. Nature Reviews. Cancer. 2014;**14**(11): 754-762

[46] Ebadi M, Mazurak VC. Evidence and mechanisms of fat depletion in cancer. Nutrients. 2014;**6**(11):5280-5297

[47] Kliewer KL et al. Adipose tissue lipolysis and energy metabolism in early cancer cachexia in mice. Cancer Biology & Therapy. 2015;**16**(6):886-897

[48] Arner P, Langin D. Lipolysis in lipid turnover, cancer cachexia, and obesity-induced insulin resistance. Trends in Endocrinology and Metabolism. 2014; **25**(5):255-262

[49] Murphy RA et al. Loss of adipose tissue and plasma phospholipids: Relationship to survival in advanced cancer patients. Clinical Nutrition. 2010;**29**(4):482-487

[50] Lampidonis AD et al. The resurgence of Hormone-Sensitive Lipase (HSL) in mammalian lipolysis. Gene. 2011;**477**(1-2):1-11

[51] Silvério R et al. Lipases and lipid droplet-associated protein expression in subcutaneous white adipose tissue of cachectic patients with cancer. Lipids in Health and Disease. 2017;**16**(1):159

[52] Laurencikiene J et al. Evidence for an important role of CIDEA in human cancer cachexia. Cancer Research. 2008; **68**(22):9247-9254

[53] Rohm M et al. An AMP-activated protein kinase-stabilizing peptide ameliorates adipose tissue wasting in cancer cachexia in mice. Nature Medicine. 2016;**22**(10):1120-1130

[54] Rosen ED, MacDougald OA. Adipocyte differentiation from the inside out. Nature Reviews. Molecular Cell Biology. 2006;**7**(12):885-896

[55] Murano I et al. Dead adipocytes, detected as crown-like structures, are prevalent in visceral fat depots of genetically obese mice. Journal of Lipid Research. 2008;**49**(7):1562-1568

[56] Guilherme A et al. Adipocyte dysfunctions linking obesity to insulin resistance and type 2 diabetes. Nature Reviews. Molecular Cell Biology. 2008; **9**(5):367-377

[57] Farmer SR. Transcriptional control of adipocyte formation. Cell Metabolism. 2006;**4**(4):263-273

[58] Alonso-Vale MI, Peres SB, Vernochet C, Farmer SR, Lima FB, et al. Adipocyte differentiation is inhibited by melatonin through the regulation of C/EBPβ: Transcriptional activity. Journal of Pineal Research. 2009;**47**(3):221-227

[59] Franco FO et al. Cancer cachexia differentially regulates visceral adipose tissue turnover. The Journal of Endocrinology. 2017

[60] Lopes MA, et al. Effects of LLC tumoral secretory products in coculture system on adipocyte differentiation. Journal Cachexia Sarcopenia Muscle. Dec 2015;**6**(4):398-509

[61] Henriques F et al. Deletion of TLR4 attenuates adipose tissue inflammation in cancer cachexia. The FASEB Journal. 2015;**29**(1 Supplement)

[62] Henriques FS et al. Atorvastatin reduces adipose tissue browning and prolong survival in cancer cachexia mice. The FASEB Journal. 2017;**31** (1 Supplement):728

[63] Olefsky JM, Glass CK. Macrophages, inflammation, and insulin resistance. Annual Review of Physiology. 2010;**72**: 219-246

[64] Streuli C. Extracellular matrix remodelling and cellular differentiation. Current Opinion in Cell Biology. 1999; **11**(5):634-640

[65] Keophiphath M et al. Macrophage-secreted factors promote a profibrotic phenotype in human preadipocytes. Molecular Endocrinology. 2009;**23**(1): 11-24

[66] Antras J et al. Decreased biosynthesis of actin and cellular fibronectin during adipose conversion of 3T3-F442A cells. Reorganization of the cytoarchitecture and extracellular matrix fibronectin. Biology of the Cell. 1989;**66**(3):247-254

[67] Spiegelman BM, Ginty CA. Fibronectin modulation of cell shape and lipogenic gene expression in 3t3-adipocytes. Cell. 1983;**35**(3, Part 2): 657-666

[68] Chun T-H et al. A pericellular collagenase directs the 3-dimensional development of white adipose tissue. Cell. 2006;**125**(3):577-591

[69] Strissel KJ et al. Adipocyte death, adipose tissue remodeling, and obesity complications. Diabetes. 2007;**56**(12): 2910-2918

[70] Wang S et al. Reduced adipogenesis after lung tumor exosomes priming in human mesenchymal stem cells via TGFbeta signaling pathway. Molecular and Cellular Biochemistry. 2017;**435** (1-2):59-66

[71] Tsoli M, Swarbrick MM, Robertson GR. Lipolytic and thermogenic depletion of adipose tissue in cancer cachexia. Seminars in Cell & Developmental Biology. 2016;**54**:68-81

[72] Tsoli M et al. Activation of thermogenesis in brown adipose tissue and dysregulated lipid metabolism associated with cancer cachexia in mice. Cancer Research. 2012;**72**(17):4372-4382

[73] Vaitkus JA, Celi FS. The role of adipose tissue in cancer-associated cachexia. Experimental Biology and Medicine (Maywood, N.J.). 2017; **242**(5):473-481

[74] Osborn O, Olefsky JM. The cellular and signaling networks linking the immune system and metabolism in disease. Nature Medicine. 2012;**18**(3): 363-374

[75] Smith BK, Conn CA, Kluger MJ. Experimental cachexia: Effects of MCA sarcoma in the Fischer rat. The American Journal of Physiology. 1993; **265**(2 Pt 2):R376-R384

Section 4

Muscle Markers: Immuno-Analysis

Chapter 8

Current Approaches in Immunoassay Methods Focus on Skeletal Muscle Proteins

Gisela Gaina

Abstract

The skeletal muscle is a complex tissue that represents most of the muscle tissue in mammals and plays a key role in health and in the body's function. It is a heterogeneous tissue whose contractile and metabolic functions depend on type, size, and quality of a large number of proteins. The multitude of proteins, the relationships that exist between them, and functional changes that occur in different muscle pathologies make their investigation to be challenged. In this chapter, current approaches in proteomic studies, its application, specific technical advice, and recent progress of the most important techniques based on antigen-antibody interactions used for the analysis of muscle proteins involved in different muscle diseases are presented.

Keywords: antigen, antibody, immunofluorescence, western bolt, antisense oligonucleotides

1. Introduction

Proteins, key players responsible for good muscle function, are the main structural and functional constituents of the muscle [1]. In the last three decades, a considerable amount of skeletal muscle proteins that present a vast variation in size, shape, abundance, and expression have been identified. Based on their localization, muscle proteins are grouped as muscle membrane (dystrophin, sarcoglycans, dysferlin, and caveolin-3), the extracellular matrix (a2-laminin and collagen VI) and from the sarcomere (telethonin, myotilin, titin, and nebulin), the muscle cytosol (calpain 3, myotubularin, TRIM32), the nucleus (emerin, lamin A/C, SMN) [2]. Mutation in the gene that encodes for specific muscle proteins is responsible for the changes in expression of most of them and lead to different forms of neuromuscular diseases.

Evaluation of the localization of a protein on tissue, its expression, cellular level, and understanding of the interactions with different other proteins can elucidate the molecular basis of muscle pathology.

The differences in terms of size (from titin with a molecular mass around 1200 kDa, nebulin 700 kDa, to caveolin 3–25 kDa) and location of each of them make their investigation difficult [3].

However, the examination of the muscle protein expression by immunostaining reactions, as well as investigation of the abundance and size using SDS gel electrophoresis by methods based on antigen-antibody interaction method, has been

IntechOpen

developed and improved over time [4]. The immunoassay methods represent a powerful tool that offers comprehensive details about the molecular alteration of different muscle pathology. Different strategies have improved over time in terms of quality and sensitivity, but each of them has the specific requirement as well as limitations.

Applications of immunoassay techniques contribute to both diagnostic and research purposes for biomarkers identification and therapeutic drug monitoring.

This chapter presents the most used techniques based on the interaction between antibodies (Ab) with a specific antigen (Ag) for muscle protein analysis, providing necessary knowledge for obtaining a good result, and provides some suggestions for the inherent problems that may be encountered in different stages.

2. Current immunoassay technologies and approaches

Immunoassay methods, technologies based on the properties of the antigen-antibody interaction, allow both qualitative and quantitative analysis of interest protein. Most important and used ***qualitative*** methods for the investigation of muscle proteins are immunohistochemistry (IHC) and fluorescent immunohistochemistry "gold standard" in muscle biopsy analysis. Their use for the evaluation of muscle proteins regarding localization, loss, accumulation, or maldistribution of different proteins involved in different myopathies has increased since the 1960s [5]. Advances in basic muscle research and development of new antibodies have increased the broad diversity of the proteins that can be analyzed. Also, technological advances in image capture, as well as improved image analysis software, now allow better interpretation and quantification of the protein expression levels both for research and for clinical pathological diagnosis. Because the successful outcome of IHC depends strictly on choosing a specific and sensitive primary antibody, and to avoid possible false-positive results, the need for an alternative method to confirm results was imperative [6]. However, for the detection and quantification of proteins in complex biological mixtures as well as for confirmation of immunohistochemistry results, western blot represents a good alternative. Western blot, also known as immunoblot, is an important quantitative method for qualitative and semiquantitative sample analysis and differential diagnosis [7].

Both immunoassay methods have various applications and have been widely used in the identification of biomarkers, diagnosis of diseases, therapeutic drug monitoring, and drug discovery, as well as characterization of protein expression and function by use of antibodies (**Table 1**).

2.1 Antigens and antibodies

2.1.1 Antigens

Antigens, also called immunogens, are any macromolecule of natural or synthetic origin such as proteins, polypeptides, polysaccharides, nucleic acids, and infectious agents that stimulate the immune system to induce an immune response (e.g., to produce a specific antibody). An antigen (Ag) molecule consists of two components: a carrier that is the largest part of the molecule and specific antigenic determinants or epitopes that are immunologic active regions of the antigens recognized by an antibody. An epitope is usually a protein segment by 5 to 8 six amino acids in length, specific for each antigen. The fixation of the antigen at the combining site of the immunoglobulin occurs based on the complementarity between epitope and paratope by multiple noncovalent bonds, van der Waals forces, hydrogen, and hydrophobic-type interactions. The rate formation of the antigen-antibody

Abbreviation	Method	Relevance to muscle proteomic studies
IHC	Immunohistochemistry	Localization of a protein Accumulation of a protein in a tissue Qualitative pattern: presence, absence, or reduction Secondary reduction
IF	Immunofluorescence	Localization of a protein Accumulation of a protein in a tissue Qualitative pattern: presence, absence, or reduction Secondary reduction
ICW	In-cell Western	Detect and quantify target proteins localized in-cell Quantitative analysis of cellular signaling pathways
WB	Western blot	Evaluation of molecular mass and abundance Differential diagnosis Level of expression
MWB	Multiplex Western blot	Localization of a protein Accumulation of a protein in a tissue Qualitative pattern: presence, absence, or reduction Secondary reduction

Table 1.
Methods and their relevance to proteomic studies.

complex depends not only on the preservation and accessibility of epitopes to the antibodies [8] as well as the concentration of the reactants but also on the mutual affinity. The alteration or destruction of epitopes by different treatment determines the reduced or abolished antigen-antibody interaction.

Because proteins are the most powerful antigens, the immunohistochemistry reaction is primarily concerned on their identification and localization as well as glycoprotein and lipoprotein compounds. The successful rate of immunohistochemistry reaction-detection of the antigen-antibody complex-depends on several factors of which the most important are: (i) the reactivity and quality of the reagents; (ii) antigen concentration in the analyzed tissue; (iii) antigen retrieval methods; and (iv) labeling conditions and methods used [9].

2.1.2 Antibodies

Antibodies also called immunoglobulins (Ig) are large Y-shaped animal glycoproteins made up of four polypeptide chains: two identical heavy chains (H) and two identical light chains (L) linked together by disulfide bonds.

While there are two main types of light chains *kappa* (κ) and *lambda* (λ) that decide the *type of immunoglobulin* molecule, for the heavy chains, 5 isotypes are found in the molecule in mammals such as humans and mice and determine the *class of immunoglobulins:* IgG, IgM, IgA, IgD, and IgE, each playing a specific role [10]. The main immunoglobulin (Ig) found in human serum that makes up about 75% of the total immunoglobulins is IgG, which is further divided into four subclasses: IgG1, IgG2, IgG3, and IgG4, each with its own biologic properties [10]. IgA can be split into two subclasses: IgA1 and IgA2 [11].

Ig is also a bifunctional molecule presenting two antigen-binding sites called Fab fragment and one complement-binding site called Fc fragment. For both heavy and light chains, the C terminal parts contain constant regions of the antibody, while the N-terminal contains variable domain through the antibody that binds to the antigen named *binding site* or *paratope*. The specificity, affinity, and antibody diversity are determined by the structure of this region [12].

The use of antibodies for the identification of protein antigen in human tissue was a revolutionary step for methods like western blotting, immunofluorescence, and immunohistochemical studies. The two most used types of antibodies in these studies are monoclonal and polyclonal antibodies.

For a successful immunohistochemistry experiment, a few things related to antibodies must be known such as whether the antibody will react with the antigen under conditions of fixation and processing system used, the immunoglobulin class to which the antibody belongs, the species in which the antibody was generated, which epitope to be targeted, type of antibody used (monoclonal or polyclonal), and analysis methods.

2.1.3 Monoclonal versus polyclonal antibodies

Besides the crucial and active role, which it plays in producing the mammal's normal immune response following the action of pathogens, antibodies are powerful tools for research and diagnostic purposes. Due to high exquisite specificity and selectivity, antibodies are an excellent tool utilized in a wide variety of therapeutic applications including detection and quantification of molecules of interest [13].

Antibodies used for research and diagnostic applications are produced by repeatedly injecting a laboratory animal (e.g., mice, rats, rabbits, and goats) with a specific antigen until it confirms the occurrence of antigen-specific antibodies in the blood serum. Generally, based on production and purification methods, antibodies can be classified into two groups: polyclonal or monoclonal.

Monoclonal antibodies (mAbs) refer to a specific antibody for a single antigenic epitope secreted from a single B-cell plasma clone. Subsequent, B-cells are fused with a myeloma cell line and grown in culture.

Polyclonal antibodies (pAbs) are produced by injecting a laboratory animal (e.g., mice, rats, rabbits, and goats) with a specific antigen and are a heterogeneous mixture of immunoglobulin molecules that are usually produced by different B-cell clone species. They can recognize and bind to several antigenic epitopes of a single antigen.

For the identification of the same antigen, both monoclonal and polyclonal antibodies are available. Choosing a primary monoclonal or polyclonal antibody for a specific target in immunoassay methods depends on the purpose of the experiment.

In the immunostaining methods based on antigen-antibody interaction, higher specificity and lower affinity of the mAbs as well as lower specificity but higher affinity of pAbs must be taken into account. To assess changes in molecular conformation, protein-protein interactions, as well as to identify the members of protein families, mAbs are more useful because of their monospecificity [14]. Polyclonal antibodies prove their usefulness in the studies they are aiming to detect variants of a particular protein of interest [15] by being able to amplify the signal of interest protein with low expression level (**Table 2**).

2.1.4 The characteristics of antibody-antigen interaction

Antigen-antibody complex (immune complex) is a bimolecular association that occurs when an antigen combines with an antibody based on the affinity of the antibody for the antigen. The reaction of the antibody-antigen is highly specific. The antibodies recognize the epitope region of the antigens and interaction between them is maintained due to exclusively noncovalent bonds such as hydrogen bonds, ionic bonds, hydrophobic interactions, and van der Waals forces [16]. The bonds are irreversible, but the antigen-antibody interaction can be affected during washing steps leading to dissociation of the complex. A lot of factors can influence the Ag-Ac complex such us temperature of incubation, pH buffer, the concentration of

	Monoclonal antibodies	Polyclonal antibodies
Advantages	Higher specificity to a single epitope Higher reproducibility Moderate sensitivity Reduced cross-reactivity Provide better results	Rapid generation and less expense costs Higher affinity–recognize and bind to many different epitopes of a single antigen Less technical skill required for production Easy to store Quicker binding to the target antigen Stable to pH and buffer conditions
Disadvantages	More time – 4-6 months to produce and develop Expensive costs for production Less tolerance to pH and buffer condition changes	Cross-reactivity due to a recognition of multiple epitopes High variability between different lots produced in different animals at different times

"Sensitivity" refers to the limit of detection of an assay for a specific target in a sample.
"Specificity" refers to the ability of a method to measure a specific target in a sample.

Table 2.
Characteristics of monoclonal and polyclonal antibodies.

antigen/antibody, and incubation time as well as the number of antigen sites per cell (zygosity) [16, 17].

2.1.5 Detection systems

The Ag-Ab complexes are not visible with standard microscopy and must be labeled. A wide range of fluorochromes is commercially available. An important property of a fluorochrome is the absorption spectrum. Fluorochromes absorb light at one wavelength [18] and emit light at a different wavelength.

The most commonly used markers for labeling are:

a. Fluorochromes: mostly used are fluorescein (absorbs blue light and emits yellow-green) and rhodamine (absorbs yellow-green light and emits deep red)

b. Enzymes for histochemical techniques (e.g., peroxidase, alkaline phosphatase, and glucose oxidase)

c. Metals for use in electron microscopy (ferritin and colloidal gold)

Most often, the sensitivity of an immunolabeling reaction depends on the selection of the detection system, which makes the choice of detection to be done carefully. The permanent attempt to obtain better results and significant advances in the biology field has led to the development of numerous methods for visualizing the antibody-antigen complex. Detection systems are classified as direct or indirect methods depending upon whether the fluorochrome is attached to the primary, secondary, or tertiary antibody. Regardless of the method chosen for labeling, both direct and indirect assay make possible distribution and precise localization of a specific protein or specific cellular components within a tissue or cell as well as the study of protein expression and function (**Table 3**).

2.1.5.1 Direct labeling

The direct labeling method is used for the detection of the point of interest in a specimen using a single primary Ab directly coupled with a reporter molecule (**Figure 1**). The method is usually shorter, involves one incubation step reaction,

	Direct	**Indirect**
Advantages	Single labeling step Short time procedure No cross-species reactivity	Great sensitivity-high amplification of the signal Production of the secondary antibodies is inexpensive
Disadvantages	Less sensitivity Lower signal Higher costs Restricted availability of direct conjugate antibodies	Double labeling steps Long procedure Extra incubation and wash steps are required High background due to endogenous activity

Table 3.
Advantage and disadvantage of immunofluorescence methods.

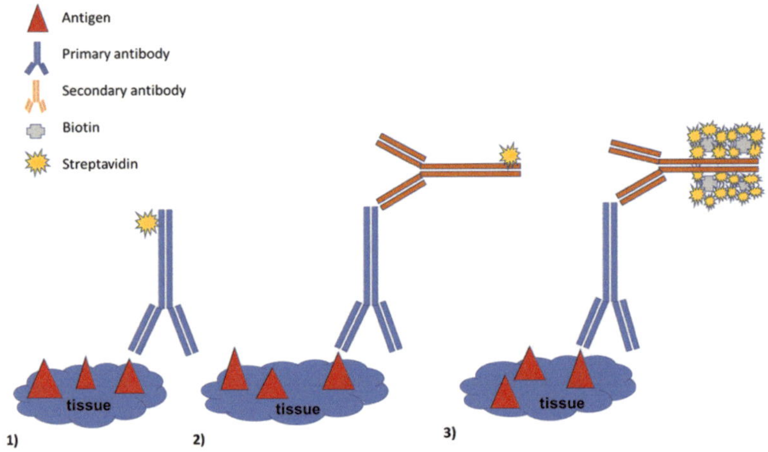

Figure 1.
Schematic representation of immunolabeling mechanisms. (1) Direct labeling; (2) Indirect labeling with secondary antibody conjugate with fluorophore; (3) indirect labeling with biotinylated secondary antibody.

and is widely used for detecting highly expressed antigens in a tissue. The advantage of the direct labeling method is the avoidance of antibody cross-reactivity or nonspecificity.

Nevertheless, direct labeling is not routinely employed in clinical and research applications due to insufficient sensitivity of the method in detecting of antigens in a tissue, the weak intensity of obtained signal, as well as optimal given preparation process of attaching fluorochrome to the antibody, which can affect antibody affinity. To date, for muscle protein analysis is used indirect labeling due to the higher sensitivity of the method.

2.1.5.2 Indirect labeling

The need to improve the sensitivity of detection methods of antigens with low expression led to the development of labeling methods in several stages, which intensifies the signal. *Coons et al.* developed in 1941 a detection method that involved two distinct steps: i) an unconjugated primary antibody incubated with tissue section that binds to the target molecule and forms an invisible antigen-antibody complex and ii) secondary antibody, which carries the fluorophores, enzyme, and biotin, recognizes, and binds to the primary antibody facilitating detection by producing a fluorescent signal or a collared substance. This method was a progress for labeling method because it allowed the amplification of the signal by increasing the number of fluorophore molecules that target the antigen. There are several advantages of the method that make it so popular: different primary antibodies can be used with the same secondary antibody, the possibility of use of secondary antibodies conjugated with different dye fluorophores, and the ability of the unconjugated primary antibody to bind to up to five secondary antibodies [19].

Thus, the indirect method is no longer used in its original form but continues to develop both the identification and production of a large number of fluorescent molecules, as well as the improvement of specific methodologies for coupling fluorochromes with antibodies.

2.1.6 Important parameters for immunomarkers

Therefore, even the immunostaining methods may seem simple in concept, there are many critical steps in performing it. To ensure success and to obtain good results, several parameters should be considered as follows:

1. ***Sample collection and tissue handling*** immediately after surgery play an important role in immunostaining. For example, [20] showed that a long time after harvesting increases degradation of calpain 3 in muscle due to *autolysis* that can affect the detection of this protein involved in a type of muscular dystrophy by immunostaining methods.

2. ***Specimen fixation*** is another important step in immunostaining. The alteration, damage, or maskings of the epitope of some antibodies against sarcolemmal membrane-associated proteins by fixation in formalin-fixed and paraffin-embedded skeletal muscle tissue have been reported [21].

3. ***Slide storage conditions***: Following freezing and until sectioning, tissue must be stored at ultra-low temperature (−80°C). The tissue architecture undergoes damage due to freeze-thaw [22]. Also, longer storage tissue sections lead to cracking as well as decrease in intensity signal levels. Furthermore, protecting the tissue slide from oxidation must be considered.

4. ***Probe size***: Depending on the purpose, the tissue section for immunostaining can vary. Previous reports have shown that the intensity of staining is indeed dependent on tissue thickness [23]. Generally, for muscle protein analysis, the recommended thickness of tissue slides is 7um for tissue cryosection [...] and 3–5 μm for FFPE sections [21].

5. ***Antibody selection (monoclonal vs polyclonal):*** This step is critical for immunostaining methods. Knowing the differences between monoclonal antibodies (that recognize a single epitope in an antigen) and polyclonal antibodies (that bind to multiple different epitopes from an antigen) for the interest target must be selected the antibodies with higher specificity. In addition, antibody concentration, working temperature, and duration of incubation affects the immunostaining reaction results and must be optimized.

6. ***Controls***: Due to insufficient specificity of antibodies or procedures used, sometimes unexpected binding and labeling can occur. To avoid confusing or inconsistent obtained results as well as to demonstrate that the detected signal

is specific for the target protein, a set of controls is required to demonstrate the reliability of labeling. Therefore, while planning immunolabeling experiment, more types of controls are essential [24] and should be included to validate the staining reaction and to show that the protocol works properly:

Tissue (antigen) controls

- *Positive control*: a tissue section, which is known to express the target protein; successful staining reactions of this control demonstrate the protocol and parameters used to detect the target protein.

- *Negative control*: a tissue section known not expressed the target protein and no response is expected. It is useful for the identification of positive errors.

- *Tissue background control*: used to avoid interpretation of autofluorescence mainly comes from mitochondria, lysosomes, for example, as positive results.

Reagent specificity controls

- *Primary antibody controls*: useful to demonstrate specific binding of the primary antibody to the antigen

- *Secondary antibody controls*: useful to show specific binding of the secondary antibody to the primary antibody

The use of antibodies in immunohistochemical microscopy-based experiments is a helpful tool for identification, localization, and expression patterns of muscle proteins. This technique is still routinely used in research to study the role of interest protein both in healthy and in pathological muscle as well as for diagnostic purposes or optimized treatment regimes and therapeutic drug monitoring. However, the use of one method only for the complete characterization of a muscle protein is not enough. The advances in the proteomic field led to the improvement of the analytical method for identification and quantification of proteins avoiding thus inconsistent and confusing immunocytochemical results.

The most used method for confirmation of immunohistochemical results as well as for the size of the protein is western blot. There are many different types and methods for Western blotting.

2.2 Western blot

The Western blot (WB) method also known as protein immunoblotting is an important analytical and quantitative technique used to identify, to separate a specific protein from a given complex biological mixture of proteins from a tissue/cellular homogenate, and to determine the amount of antigens (proteins) in reaction with a specific antibody. Initially, the proteins are electrophoretically separated into a polyacrylamide (PAA) gel electrophoresis. The most widely used technique for large scale protein from a mixture of proteins is sodium dodecyl sulfate polyacrylamide gel electrophoresis (SDS-PAGE) due to the possibility of separation of proteins based on their molecular weight under the action of electric current after linearization of the proteins. Special gradient gels are used to separate proteins with a wide variety of molecular mass. After the separation, the proteins are electrophoretically transferred on a solid substrate such as nitrocellulose, PVDF, or cationic nylon membrane. To suppress nonspecific adsorption of the antibodies, the unreacted nonreactive binding sites on the membrane are blocked, which causes immobilized proteins to react specifically with monoclonal or polyclonal antibodies. Primary antibodies can be applied single when we target the expression of a single protein or in a cocktail when a simultaneous analysis of several proteins is necessary. Commonly used conjugates for secondary antibody are color, radioactivity, enzyme, as well as biotin. Antigen-antibody complexes are radiographically, chromogenically, or chemiluminescently visualized.

With this technique, the size (molecular mass) and abundance of the interest protein are evaluated by comparing with the control. The need for the analyses of multiple different target proteins simultaneously led to great improvement of sample separation resolution.

2.2.1 Multiplex Western blot

The multiplex detection methods have improved over the past few decades and have led from the analysis of a single specific protein to the detection of simultaneously multiple target proteins with a different molecular weight in a complex of cellular homogenates [25, 26]. For the simultaneous analysis of muscle proteins, a biphasic polyacrylamide gel system with different concentrations is used. With this system, the separation of muscle proteins is done based on their molecular weight (proteins with molecular weight more than 200 kDa are separate in the top part of the gel, e.g., dystrophin, while the smaller proteins under 150 kDa, e.g., calpain 3 in the bottom). Highlighting of proteins is achieved with a cocktail of specific primary antibodies.

Simultaneous analysis of multiple proteins involved in different muscle pathologies revolutionized the medical diagnosis, reduced cost and time for analysis, and improved differential diagnosis in muscle pathology.

2.2.2 In-cell western (ICW) assay

ICW also known as in-cell ELISA (ICE) is quite a novel quantitative immunofluorescence-based technique suitable for the detection of protein levels and signaling events performed in cell culture grown on microplate format [27]. It is an extremely sensitive method and accurately quantifies, which can detect two targets normally labeled with specific primary antibody followed by incubation with secondary antibody fluorescent conjugated with spectrally distinct dyes and quantification of the signals from fluorophores conjugated at different wavelengths on two detection channels. The technique allowed the quantification of proteins directly in cell culture [28]. The accuracy of quantification is increased by normalization due to adjustments of the cell number in wells. This technique is also useful in the study of the drug effect on multiple points.

2.3 Data acquisition and image analysis

For the evaluation of antigen expression by ***immunohistochemistry*** (IHC) and fluorescent IHC, a comprehensive microscopic evaluation of the signal beside specimen preparation and staining protocol depends also on the type of microscope used for acquisition of imagines.

2.3.1 Light microscopy

In bright-field microscopic IHC, antigen expression in muscle is visualized by a combination of a secondary antibody with an enzyme and utilizing a colorimetric substrate that produces a colored reaction product detected by light microscopy. The most widely used enzymes in chromogenic detection are the horseradish peroxidase (HRP) and alkaline phosphatase (AP), which convert 3,3′ diaminobenzidine (DAB) and 3-amino-9-ethylcarbazole (AEC) into brown and red end color, respectively. The reaction product is stable for a long time and visualization can be done any time. Multiple labeling is also possible since different substrates are now available for the same enzyme, but low contrast and resolution for samples is a limitation of bright-field microscopy.

2.3.2 Fluorescent microscopy

Although a lot of muscle proteins can be detected by IHC with peroxidase and visualized in light microscope field, the labeling of antigen with an antibody coupled with a fluorochrome permitted a superior viewing of many muscle proteins and offered significant advantages in accurately identifying certain protein localizations and distinguishing subtle difference in protein expression patterns. The major advantage of fluorescence microscopy is given with the possibility of uses of different dyes that span the entire visible spectrum to track different target proteins. Also, the identification of the interaction between proteins as well as their specific localization is better observed by colocalization when the color of two or more different fluorescent dyes used appears changed as a result of the overlap. One limitation of fluorescence microscopy remains the loss of the fluorescent capacity *of* fluorophores (photobleaching) due to many cycles of excitation and emission [29]. Therefore, it is recommended to keep the samples marked in dark when not in use and use a mounting medium with an antifade agent to fix this problem.

In combination with laser confocal scanning microscope, fluorescence is preferable to evaluate the degree of colocalization and relative quantitation of proteins by specialized software [30].

2.3.3 Confocal microscopy

Confocal microscopy is a powerful laser scanning method used for the imaging analysis of fluorescently labeled specimens.

The technique provides high-resolution detailed information about the 3D structure of the tissue sections and cells allowing detailed analyses and measurements of double, triple, and even quadruple stained sections [31, 32]. This is possible due to labeling of secondary antibodies with different fluorophores, which emitted on a different wavelength.

The z-stack mode permitted three-dimensional reconstructions of optical sectioning useful for the study of the relationship between stained structures.

2.4 Qualitative and quantitative analysis of proteins

Muscle pathology usually involves changes in level and protein expression. For muscle protein diagnosis, a good knowledge is required of the expected cellular location: at sarcolemmal *level (dystrophin, sarcoglycans, dysferlin, and caveolin-3),* in the cytoplasm (*calpain 3 and TRIM32),* and nuclear (*emerin and lamin A/C)* level. The initial choice of a qualitative or quantitative method for protein analysis depends on the availability of the methods and the specificity of the antibodies. It should be mentioned that the results obtained by one method have to be confirmed by another; see **Table 4**.

Disease	Gene	Primary protein defect	Secondary changes	Localization level	Clue to diagnosis IF/WB	References
DMD	DMD	Dystrophin	Utrophin upregulated Sarcoglycans reduced/absent Dystroglycan reduced/absent nNOS absent	Subsarcolemmal	Absent, reduced/absent	[2, 43]
BMD	DMD	Dystrophin	Utrophin upregulated Sarcoglycans reduced/absent Dystroglycan reduced/absent nNOS reduced/absent	Subsarcolemmal	Reduced/Reduced in size/amount Absence of at least one antibody	[2, 43]
DMD/ BMD carriers	DMD	Dystrophin	Utrophin upregulated Sarcoglycans reduced/absent Dystroglycan reduced/absent nNOS may be absent	Subsarcolemmal	Mosaic pattern/ Reduced in size	[2, 43]
LGMD2C-F	SGCG SGCA SGCB SGCD	Sarcoglycans	Reduction of other sarcoglycan B-Dystroglycan reported Possible reduction Dystrophin Loss of nNOS reported	Sarcolemmal	Variably reduction/absent	[44]
LGMD2B	DYSF	Dysferlin	Caveolin-3 reduced	Sarcolemmal	Absent/reduced	[45]
LGMD1C	CAV3	Caveolin-3	Dysferlin reduced	Sarcolemmal	Absent/reduced	[45]
CMD	LAMA2	Laminin α2	Deficiencies of laminin β2, α-dystroglycan and integrin α7	Extracellular matrix	Completely or partially absent	[46]
Bethlem or Ullrich myopathy		Collagen VI	Deficiency of laminin β1 chain	Extracellular matrix	Reduced/absent	[47]
LGMD2G	TCAP	Telethonin	—	Sarcomere	Reduced/absent	[44, 48]
LGMD2J	TTN	Titin	Calpain 3 reduction	Sarcomere	Absence of calpain 3/loss of C-terminal fragments of titin results in the reduction of higher molecular weight	[48, 49]

Disease	Gene	Primary protein defect	Secondary changes	Localization level	Clue to diagnosis	References
					IF/WB	
LGMD1A	MYOT	Myotilin	Secondary laminin γ reduction	Sarcomere	Protein aggregates	[45, 48]
LGMD 1E	DES	Desmin	Myotilin, αB-crystallin, VCP cytoplasmic aggregates	Exosarcomeric	Desmin cytoplasmic aggregates	[50]
LGMD2A	CAPN3	Calpain 3	Dysferlin reduced	Cytosol	Labeling may be absent or reduced on sections/ Calpain 3 bands may be variably reduced	[44, 50]
LGMD 2H	TRIM32	Trim 32	—	Cytosol	Reduced expression/reduced level	[51]
Emery-Dreifuss MD	EMD	Emerin	—	Nucleus	Absent /reduced level	[52]
LGMD1B	LMNA	Lamin A/C	Laminin β1 reduction	Nucleus	Lamin A/C normally expressed	[44]

Table 4.
The skeletal muscle-specific proteins.

2.4.1 Analysis of immunohistochemical expression

Immunohistochemical staining is useful for the identification of the interest markers and provides valuable information about their distribution, localization, and expression in a tissue or cell. The obtained results should be interpreted by comparing a given protein pattern in normal and affected tissue in the presence of the controls of the reaction mentioned above. The evaluation of muscle protein expression is generally made in a qualitative manner based on their presence, absence, or variable reduction. Depending on the manifestation of the protein identified by immunohistochemical analysis, the progression of different neuromuscular disorders can be evaluated.

The usefulness of the method lies in the identification of the primary protein abnormalities in recessive diseases [33] and secondary reduction of interconnected proteins. For example, in the primary reduction of dystrophin, protein involved in Duchenne/Becker muscular dystrophy and secondary reduction of sarcoglycans [22] and of cytosolic calpain 3 [25] was reported. The reduction of dysferlin also is accompanied by the reduction of calpain 3.

Because of the currently limited access to next-generation sequencing that permitted analysis of simultaneous gene involved in a different type of muscular dystrophies, analysis of muscle biopsy by the immunoassay method is still a helpful method in many laboratories. Different computational methods developed in the last decade for quantitative immunohistochemical (IHC) image analysis of proteins have begun to be increasingly used.

How much protein is needed for proper muscle function has always been a problem that researchers have been trying to solve. Immunohistochemistry quantification of dystrophin proteins involved in the most severe type of muscular dystrophy has come to the attention of researchers with the improvement of image analysis software. Thus, L. E. Taylor [34] and Antony K [35] developed a method of image analysis that allows immunofluorescent quantification of dystrophin expression in sections that proved to be robust and reliable method of biomarker detection. These methodologies contribute to and improve the final diagnosis and especially are used in the analysis of a protein after a drug and specific treatment.

However, increasing the contradictory results obtained by this method and reported in the literature, a final and accurate diagnosis requires the confirmation of the results by another quantitative method, such as Western blot.

2.4.2 Interpretation of Western blot results

Western blot methods have become very popular in diagnostic laboratories due to the ability to analyze several proteins simultaneously avoiding thus the preservation of large portions from an affected muscle [36, 37].

The rapid evolution of the quantitative methods has improved over time from the analysis of one protein to simultaneous analysis by SDS gel electrophoresis of several proteins with different molecular mass. Multiplex Western blot technique developed by Andreson [25] represented a significant improvement in muscle protein analysis regarding efficiency and cost. By this method, can be evaluated the molecular mass (normal or reduced size) and abundance of the proteins by comparison with controls and the use of quantitative software analysis.

Western blot has greater importance in muscle protein analysis especially in the differential diagnosis of muscular dystrophies (distinguish between DMD and BMD patients); for analysis of calpain 3 in it, there are no antibodies available for immunoreactions on the sections.

2.4.2.1 Applications, utility, and importance of immunoassay methods

The immunohistochemical reactions are widely used both in the investigation and in the pathological diagnosis evaluations of diseases based on the antigen-antibody reactions. The identification and localization of an antigen in a tissue or cells can provide valuable information, which otherwise could not be obtained by other methods. Because most of the antigens are usually proteins, identifications of abnormal expression patterns of proteins in diseased tissue are also helpful in differential diagnosis and detecting primary protein defect involved in the pathology. Besides the identification of a primary defect and protein expression chances, immunohistochemistry is also helpful in the pharmaceutical analysis area for drug discovery and therapeutic drug monitoring.

2.4.3 Therapeutic drug monitoring

With knowledge advance in disease pathogenesis and the development of novel therapeutic strategies, there is a need to generate new treatments for specific neurodegenerative disorders. In the last few years, the use of antisense oligonucleotides (AOs) has increased interest in the treatment of the different types of muscular dystrophy such as Duchenne muscular dystrophy.

Antisense oligonucleotides, a new class of the synthetic single-stranded molecules of nucleic acids as RNA or DNA, have been reported that modulate the gene expression and splicing process interacting with specific gene transcripts through a variety of mechanisms and restore the expression of functional protein [38, 39].

The quantification of protein levels after treatment will be monitored by the immunoassay method after administration of a specific dose of specific AOs. Positive results obtained in clinical trials with AOs for different diseases increase the interest in the clinical application of antisense strategies. However, more clinical trials and more data are necessary to make this strategy clinically available.

Difficulties, limitations, and disadvantages of immunostaining methods and troubleshooting are discussed further.

a. Difficulties

Although it is a simple technique, many parameters must be considered and optimized before. The quality of the results depends on these critical factors that begin from handling the specimen to tissue fixation procedure, detection system, staining protocol, antibody selection, and sensitivity. Also, the ability, experience, the rigor of execution of the researcher in performing the reaction, and good knowledge and understanding of the methodology and morphological changes in a specific pathology are necessary for an accurate overview of protein expression and interpretation of the results to avoid misinterpretation [40, 41].

b. Limitations

One of the major limitations of the immunoassays is antibody specificity. The nonspecific binding occurs when an antibody attaches to a cell without a specific epitope for that antibody. Several reasons are responsible for nonspecific binding such as higher concentrations of the antibody and binding by *Fc* segment contained by most of the antibodies to *Fc* receptors of immune cells (e.g., neutrophils, monocytes, and macrophages). The decrease of antibody concentration and the use of an Fc blocking reagent will minimize the nonspecific binding [42].

The nonspecific binding of the antibody can cause errors and confusion in the interpretation of the obtained results. To eliminate the interpretational error, antibody specificity controls must be introduced into the reaction to obtain reliable staining [43].

A limited number of available antibodies for muscle protein represent another limitation of this method as well as the choice of an insufficiently specific and sensitive antibody for the desired target. Often, the resolution microscope is a limitation in determining the proper intensity of the signal of the target antigen as well as subtle differences in protein expression level. This problem can be solved by confocal laser scanning microscopes, which offer a possibility to overcome this problem by providing images in the highest resolutions and can predict accurately variation in staining intensity. It must be considered that the image quality degrades over time.

c. Disadvantages

Both direct and indirect immunofluorescence methods have some advantages and disadvantages, which are presented in **Table 3**. Besides, these immunostaining methods have some disadvantages among which we mention the lack of worldwide standardized protocol that determines the introduction of a level of subjectivity both in the working procedure and in the interpretation of the obtained results. Once again, the experience of the personnel who work this technique is crucial.

3. Conclusion

Immunoassay methods are a powerful tool for characterization of the proteins regarding localization, understanding the quantitative and qualitative characteristics, as well as interactions of proteins at the cellular level. The progress of the development of the specific antibodies, as well as improvement in image analysis software, leads to more sensitive and specific immunoassay methods used for characterization of the proteins. However, a combination of qualitative and quantitative methodologies offers great value for the results. Further advances in immunoassay methods will lead to a better understanding of the functional role of proteins.

Acknowledgements

This work was funded by Ministry of Research and Innovation in Romania, under Program 1-The Improvement of the National System of Research and Development, Subprogram 1.2-Institutional Excellence-Projects of Excellence Funding in RDI, Contract No. 7PFE/16.10.2018; National Program 31 N/2016/PN 16.22.02.05; COST Action Delivery of Antisense Oligonucleotides.

Conflict of interest

The author declares that there are no competing interests.

Acronyms and Abbreviations

Ag	antigen
AOs	antisense oligonucleotides
BMD	Becker muscular dystrophy
DMD	Duchenne muscular dystrophy
CMD	congenital muscular dystrophy
ICW	in-cell Western
IF	immunofluorescence
Ig	immunoglobulins
IHC	immunohistochemistry
LGMD	limb-girdle muscular dystrophy
MAbs	monoclonal antibodies
MEB	muscle-eye-brain disease
PAbs	polyclonal antibodies
WB	Western blot

Author details

Gisela Gaina[1,2]

1 Victor Babes National Institute of Pathology, Bucharest, Romania

2 Department of Biochemistry and Molecular Biology, University of Bucharest, Bucharest, Romania

*Address all correspondence to: giselagaina@yahoo.com; gisela.gaina@ivb.ro

References

[1] Gillies AR, Lieber RL. Structure and function of the skeletal muscle extracellular matrix. Muscle & Nerve. 2011;**44**(3):318-331

[2] Vainzof M, Zatz M. Protein defects in neuromuscular diseases. Brazilian Journal of Medical and Biological Research. 2003;**36**(5):543-555

[3] Whiteley G, Collins RF, Kitmitto A. Characterization of the molecular architecture of human caveolin-3 and interaction with the skeletal muscle ryanodine receptor. The Journal of Biological Chemistry. 2012;**287**(48):40302-40316

[4] Kinoshita E, Kinoshita-Kikuta E, Koike T. The cutting edge of affinity electrophoresis technology. Proteome. 2015;**3**(1):42-55

[5] Vogel H, Zamecnik J. Diagnostic Immunohistology of muscle diseases. Journal of Neuropathology and Experimental Neurology. 2005;**64**(3):181-193

[6] Ohlendieck K. Skeletal muscle proteomics: Current approaches, technical challenges and emerging techniques. Skeletal Muscle. 2011;**1**:6

[7] Rabilloud T, Chevallet M, Luche S, Lelong C. Two-dimensional gel electrophoresis in proteomics: Past, present and future. Journal of Proteomics. 2011;**74**(10):1829-1841

[8] Vidarsson G, Dekkers G, Rispens T. IgG subclasses and allotypes: From structure to effector functions. Frontiers in Immunology. 2014;**5**:520

[9] Garsha K. Concepts, considerations and control experiments for immunolabeling. In: Biological Electron Microscopy. Springer Humana Press; 1992. pp. 309-320

[10] Schroeder Harry W, Cavacini L. Structure and function of immunoglobulins. The Journal of Allergy and Clinical Immunology. 2010;**125**(2 0 2):S41-S52

[11] Knoppova B et al. The origin and activities of IgA1-containing immune complexes in IgA nephropathy. Frontiers in Immunology. 2016;**7**:117

[12] Igawa T, Tsunoda H, Kuramochi T, Sampei Z, Ishii S, Hattori K. Engineering the variable region of therapeutic IgG antibodies. MAbs. 2011;**3**(3):243-252

[13] Reichert JM. Antibodies to watch in 2016. MAbs. 2016;**8**:197-204

[14] Lipman NS, Jackson LR, Trudel LJ, Weis-Garcia F. Monoclonal versus polyclonal antibodies: Distinguishing characteristics, applications, and information resources. ILAR Journal. 2005;**46**(3):258-268

[15] Wagih O, Galardini M, Busby BP, Memon D, Typas A, Beltrao P. A resource of variant effect predictions of single nucleotide variants in model organisms. Molecular Systems Biology. 2018;**14**:e8430

[16] Reverberi R. Lorenzo Reverberi factors affecting the antigen-antibody reaction. Blood Transfusion. 2007;**5**(4):227-240

[17] Foote J, Eisen HN. Kinetic and affinity limits on antibodies produced during immune responses. Proceedings of the National Academy of Sciences of the United States of America. 1995;**92**:1254-1256

[18] König K. Multiphoton microscopy in life sciences. Journal of Microscopy. 2000;**200**:82-104

[19] Sternberger LA, Joseph SA. The unlabeled antibody method.

Contrasting color staining of paired pituitary hormones without antibody removal. The Journal of Histochemistry and Cytochemistry. 1979;**27**:1424-1429

[20] Richard Charlton Henderson M, Richards J, Hudson J, Straub V, Bushby K. Barresi. Immunohistochemical analysis of calpain 3: Advantages and limitations in diagnosing LGMD2A. Neuromuscular Disorders. 2009;**19**(7):449-457

[21] Suriyonplengsaeng C, Dejthevaporn C, Khongkhatithum C, Sanpapant S, Tubthong N, Pinpradap K, et al. Immunohistochemistry of sarcolemmal membrane-associated proteins in formalin-fixed and paraffin-embedded skeletal muscle tissue: A promising tool for the diagnostic evaluation of common muscular dystrophies. Diagnostic Pathology. 2017;**12**:19

[22] Kumar A, Accorsi A, Rhee Y, Girgenrath M. Do's and Don'ts in the preparation of muscle Cryosections for histological analysis. Journal of Visualized Experiments. 2015;**99**:52793

[23] McCampbell AS, Raghunathan V, Tom-Moy M, Workman RK, Haven R, Ben-Dor A, et al. Tissue thickness effects on Immunohistochemical staining intensity of markers of cancer. Applied Immunohistochemistry & Molecular Morphology. 2017;**27**(5):345-355

[24] Burry RW. Controls for immunocytochemistry an update. The Journal of Histochemistry and Cytochemistry. 2011;**59**(1):6-12

[25] Anderson LV, Davison K. Multiplex Western blotting system for the analysis of muscular dystrophy proteins. The American Journal of Pathology. 1999;**154**(4):1017-1022

[26] Gaina G, Manole E, Bordea C, Ionica E. Analysis of muscle Calpain3 in LGMDA2A, Studia Universitatis" Vasile Goldis" Arad. Life Sciences Series. 2008;**18**:181-186

[27] Boveia V, Schutz-Geschwender A. Quantitative analysis of signal transduction with In-cell Western immunofluorescence assays. Detection of Blotted Proteins. 2015;**1314**:115-130

[28] Ruiz-Del-Yerro E, Garcia-Jimenez I, Mamchaoui K, Arechavala-Gomeza V. Myoblots: Dystrophin quantification by in-cell western assay for a streamlined development of DMD treatments. Neuropathology and Applied Neurobiology. 2018;**44**(5):463-473

[29] Tsurui H, Nishimura H, Hattori S, Hirose S, Okumura K, Shirai T. Seven-color fluorescence imaging of tissue samples based on Fourier spectroscopy and singular value decomposition. The Journal of Histochemistry and Cytochemistry. 2000;**48**:653-662

[30] Moser B, Hochreiter B, Herbst R, Schmid JA. Fluorescence colocalization microscopy analysis can be improved by combining object-recognition with pixel-intensity-correlation. Biotechnology Journal. Jan 2017;**12**(1):1600332

[31] Shrestha D, Jenei A, Nagy P, Vereb G, Szollosi J. Understanding FRET as a research tool for cellular studies. International Journal of Molecular Sciences. 2015;**16**:6718-6756

[32] Wuxiuer D, Zhu Y, Ogaeri T, Mizuki K, Kashiwa Y, Nishi K, et al. Development of pathological diagnostics of human kidney cancer by multiple staining using new fluorescent Fluolid dyes. BioMed Research International. 2014;**2014**:437871

[33] Barresi R. From proteins to genes: Immunoanalysis in the diagnosis of muscular dystrophies. Skeletal Muscle. 2011;**1**:24

[34] Taylor LE, Kaminoh YJ, Rodesch CK, Flanigan KM. Quantification of

dystrophin immunofluorescence in dystrophinopathy muscle specimens. Neuropathology and Applied Neurobiology. 2012;**38**(6):591-601

[35] Anthony K et al. Dystrophin quantification biological and translational research implications. Neurology. 2014;**83**(22):2062-2069

[36] Mahmood T, Yang P-C. Western blot: Technique, theory, and trouble shooting. North American Journal of Medical Sciences. 2012;**4**(9):429-434

[37] Aebersold R, Burlingame AL, Bradshaw RA. Western blots versus selected reaction monitoring assays: Time to turn the tables? Molecular and Cellular Proteomics. 2013;**12**:2381-2382

[38] Aartsma-Rus A, Fokkema I, Verschuuren J, et al. Theoretic applicability of antisense-mediated exon skipping for Duchenne muscular dystrophy mutations. Human Mutation. 2009;**30**:293-299

[39] Martinovich KM , Shaw NC, Kicic A, Schultz A, Fletcher S, Wilton SD et al. The potential of antisense oligonucleotide therapies for inherited childhood lung diseases

[40] de Matos LL, Trufelli DC, de Matos MGL, da Silva Pinhal MA. Immunohistochemistry as an important tool in biomarkers detection and clinical practice. Biomark Insights; 9 Feb 2010;**5**:9-20

[41] Yaziji H, Barry T. Diagnostic immunohistochemistry: What can go wrong? Advances in Anatomic Pathology. 2006;**13**:238-246

[42] Hulspas R, O'Gorman MRG, Wood BL, Gratama JW, Sutherland DR. Considerations for the control of background fluorescence in clinical flow cytometry. Cytometry, Part B: Clinical Cytometry. 2009;**76B**:355-364

[43] Rhodes KJ, Trimmer JS. Antibodies as valuable neuroscience research tools versus reagents of mass distraction. The Journal of Neuroscience. 2006;**26**(31):8017-8020

[44] Schnell FL, Wilton S. Challenges of interpreting dystrophin content by Western blot. US Neurology. 2019;**15**(1):40-46

[45] Diniz G, Eryaşar G, Türe S, et al. A regional panorama of dysferlinopathies. Turkish Journal of Pathology. 2012;**28**:259-265

[46] Toda T, Kobayashi K, Takeda S, et al. Fukuyamatype congenital muscular dystrophy (FCMD) and alphadystroglycanopathy. Congenital Anomalies. 2003;**43**:97-104

[47] Brockington M, Blake DJ, Prandini P, et al. Mutations in the fukutin-related protein gene (FKRP) cause a form of congenital muscular dystrophy with secondary laminin alpha2 deficiency and abnormal glycosylation of alpha-dystroglycan. American Journal of Human Genetics. 2001;**69**:1198-1209

[48] Vainzof M et al. Telethonin protein expression in neuromuscular disorders. Biochimica et Biophysica Acta. 2002;**1588**(1):33-40

[49] Harris E, Töpf A, Vihola A, Evilä A, Barresi R, Hudson J, et al. A "second truncation" in TTN causes early onset recessive muscular dystrophy. Neuromuscular Disorders. 2017;**27**(11):1009-1017

[50] Messina DN, Speer MC, Pericak-Vance MA, McNally EM. Linkage of familial dilated cardiomyopathy with conduction defect and muscular dystrophy to chromosome 6q23. American Journal of Human Genetics. 2007;**61**(4):909-917

[51] Servián-Morilla E et al. Altered myogenesis and premature senescence

underlie human TRIM32-related myopathy. Acta Neuropathologica Communications. 2019;7:30

[52] Manilal S, Recan D, Sewry CA, Hoeltzenbein M, Llense S, Leturcq F, et al. Mutations in Emery-Dreifuss muscular dystrophy and their effects on emerin protein expression. Human Molecular Genetics. 1998;7:855-864